Benedict Nnolim: Worked Examples in Hea

WORKED EXAMPLES

IN

HEAT CONDUCTION

Benedict Nnolim, FNSChE, FNSE

Formerly of the Department of Chemical Engineering,
Nnamdi Azikiwe University, Awka.
Anambra State, Nigeria

First published May 21, 2014

Copyright 2014 by Benedict Nnolim

Paperback edition

ISBN 978-1-906914-60-8

Other Engineering Books by Ben Nnolim Books

Title	Paperback	Hardcover
Applied Heat Transfer Volume One: Heat Conduction in Solids	ISBN: 9781906914752	
Applied Heat Transfer Volume Two: Heat Convection in Fluids	ISBN: 9781906914226	ISBN: 9781906914714
Fundamentals of Mass Transfer	ISBN: 9781906914011	ISBN: 9781906914721
Worked Examples in Mass Transfer	ISBN: 9781906914462	ISBN: 9781906914691
Worked Examples in Mass Transfer, 3rd Edition	ISBN: 9781906914998	
Worked Examples in Chemical Reaction Engineering	ISBN: 9781906914196	
Worked Examples in Chemical Reaction Engineering	ISBN: 9781906914981	

Ben Nnolim Books,
7 Sandway Path,
St. Mary Cray, Orpington, Kent
BR5 3TS, UK
Email: benedictnnolim@aol.com

Table of Contents

PREFACE

Available technical literature, especially, in heat transfer, is, largely, driven by the competition to present the latest information and state of knowledge on the subject with the result that their treatment of this subject is so advanced that it becomes incomprehensible and, often, not helpful to a beginner or student in the subject. Most of these modern text take, as given, that fundamental definitions and concepts of the subject matter are, already, well known and understood.

The reality is quite different with the consequence that most beginners and students of the subject find it difficult to lock in to the state of knowledge in the subject which their job, research or academic activity requires.

This book is built up on the premises that

i. it is necessary, for all practitioners in the subject, to have a thorough and kaleidoscopic understanding of all the essential concepts, theories, methods, calculations, etc., in the subject of heat transfer as a science and heat conduction as a branch of that science

ii. it is necessary to develop the ability to formulate and/or solve problems, in as many areas as possible, in heat conduction.

iii. a question and answer format, rather than a full prose style, may be more helpful for some but not for others.

iv. that this format is an alternative to the full prose edition of this book which is, already, available.

Both books owe a lot to a lifetime of contact with several standard textbooks on the subject such as those by Kay J. M. (*An Introduction to Fluid Mechanics and Heat Transfer*), Kays W. M. and London A. L. (*Compact Heat Exchangers*), Coulson J. M. and Richardson J. F. (*Chemical Engineering, Volume 1*), Carslaw H. S. and Jaeger J. C. (*Conduction of Heat in Solids*), Welty J. R. (*Engineering Heat*

Transfer), Welty J. R., Wicks C. E., and Wilson R. E. (*Fundamentals of Momentum, Heat and Mass Transfer*).

Much is, also, owed to several handbooks in physics, chemistry and engineering such as the *Chemical Engineers Handbook* by Perry R,H. and Green D, *CRC Handbook of Physics and Chemistry*, several articles in the chemical engineering related magazines and journals of the American Institute of Chemical Engineers, the British Institution of Chemical Engineers, the American Chemical Society and internet sites such as those of Wikipedia, M.I.T, classroom-energy.org, engineeringtoolbox.com, etc.

Finally, it is hoped that this question and answer edition and its alternative edition in straight prose both advance the ability of engineers to solve problems in engineering heat conduction.

Benedict N. Nnolirn
May 21, 2014

CHAPTER ONE
FUNDAMENTAL DEFINITIONS AND CONCEPTS

Example 1.00: What is Heat?

Answer: Heat is energy in transition as a result of a temperature difference. This transition, always, occurs, except by the intervention of other forms of energy, from a region of higher temperature (higher internal energy) to one of a lower temperature (lower internal energy).

Example 1.01: What does it mean that heat is energy in transition?

Answer: This means that heat, as thermal energy, cannot be stored in a system and is defined only as it moves or is moved (transferred or transported) from one part of a system to another, from one system to another or from a system to its surroundings, etc. This means that heat content is not defined. Other forms of energy, however, such as internal energy, potential energy, chemical energy, etc., can be stored in a system

Example 1.02: What is Energy?

Answer: Energy is the driving force of the known universe and is one of the two properties which describe the state of matter or an object. The other one is mass and both are interchangeable. Energy has two basic properties which are a). that it cannot be created or destroyed (the law of conservation of energy) and b). that it can only be transformed into another kind of energy or to mass.

Energy is, also, defined as the capacity of a system to do work. It is this definition which gives rise to its unit of measurement, in the S. I. system of units, the Joule, which is the work done

1

when a unit of force, the Newton, moves a distance of one metre. That is

$$1\,Joule = 1\,Newton.\,metre \qquad (1.01)$$

Example 1.03: What, in fact, is the universe?

Answer: The universe is, commonly, defined as the totality of existence which includes planets, stars, galaxies, the contents of intergalactic space, the smallest subatomic particles, and all matter and energy. Other terms which describe the same concept are the *cosmos*, the *world*, *reality*, and *nature*.

The observable universe is about 46 billion light years in radius with a density of 1.67×10^{-27} kg/m^3. Scientific observation of the universe has led to inferences of its earlier stages. These observations suggest that the universe has been governed by the same physical laws and constants throughout most of its extent and history.

Various **multiverse hypotheses** have suggested that the Universe might be one among many or even an infinite number of universes. Adherents of the **String Theory** of the universe, for example, think that the observable universe is just one of the 10^{500} universes in a grander universe.

We, sometimes, define our terrestrial environment, the planet Earth, as a universe because this definition is the most convenient for some analysis we wish to make.

There is the universe of classical thermodynamics which is any system with prescribed boundaries which we define and wish to analyse.

2

We define these types of universes because we find them convenient for the application of the type of physics most applicable to them. For example, Einstein physics with its theory of relativity, is more suited to the analysis of events in the cosmos, while Newtonian physics is better suited to terrestrial events. This does not mean that Einstein or Newtonian physics do not apply in all universes. It is just that each type of physics is easier to use and apply to the events which take place in the universe to which it is applied.

Example 1.04: What are the main concerns of scientific enquiry about the universe?

Answer: The ultimate questions and concerns about the universe, in addition to the utilitarian or phenomenological curiosities of mankind, continue to be a) the origin of the universe and b) the ultimate fate of the universe.

While the Big Bang theory is the prevailing cosmological model that describes the early development of the universe, calculated to have begun 13.798 ± 0.037 billion years ago, there are many other competing theories about the ultimate fate of the universe. Physicists remain unsure about what, if anything, preceded the Big Bang.

Observations of supernovae have shown, however, that the universe is expanding at an accelerating rate.

Example 1.05: What other definitions of energy are in use?

Answer: Energy has interesting definitions in classical mechanics, in classical and statistical thermodynamics, in quantum mechanics and in the theory of relativity. Only simple and general definitions of these will be given here.

In classical mechanics, the total energy of a system (kinetic and potential) is, sometimes, referred to as its Hamiltonian and the classical equations of motion can be entirely written in terms of the Hamiltonian. These equations are found to be, remarkably, similar to those obtained in non-relativist quantum mechanics.

The Lagrangian is defined, in classical mechanics, as the kinetic energy minus the potential energy. The Lagrangian formulation is said to be mathematically more convenient than the Hamiltonian for non-conservative systems such as those with friction.

An important element in the applications of the Hamiltonian and Lagrangian concepts of energy is Noether's first theorem which asserts that any differentiable symmetry of the action of a physical system has a corresponding conservation law. Its usefulness is as a fundamental tool of modern theoretical physics and the calculus of variations and in generalising the formulations of the constants of motion in Lagrangian and Hamiltonian mechanics (1).

In thermodynamics, internal energy is defined as the sum of all microscopic forms of energy of a system such as presented by its molecular structure, its crystal structure and other geometric arrangements as well as by the motion of the particles in the form of kinetic energy. This energy is also governed by the first law of thermodynamics.

The principle of equi-partition of energy states that the total energy of a system is equally split among all available degrees of freedom (entropy law)

Example 1.06: Is energy the same or different for all universes?

4

Answer: Energy is the same, by nature, for all universes but, like its physics, is presented a bit differently for each universe which may give the wrong impression that it is different for each universe.

Example 1.07: How does energy manifest itself in each universe?

Answer: All energy manifests itself, everywhere, as a proportionate amount of mass. For example, in the cosmos, Einstein physics states that mass and energy are interchangeable through the equation

$$E = mc^2 \qquad\qquad (1.02)$$

where **m** is mass and **c** is the velocity of light. The same law is valid on earth but when applied to the earth, calculations will show that adding, for example, 25 kilowatt-hours or 90 mega Joules of energy to a system will increase its mass by only 1 microgram. This makes the Einstein definition of energy not very useful for terrestrial applications for which Newton physics, which defines energy as the product of force and the distance moved by the force, is more practical.

Example 1.08: Do we know how much energy there is in the universe?

Answer: Yes and no. All we have is an estimate based on known and measurable phenomena. On the basis of the **zero-energy universe** hypothesis, the total amount of energy in the universe is exactly zero: its amount of positive energy in the form of matter being exactly cancelled out by its negative energy in the form of gravity.

Example 1.09: If the total energy of the universe is zero why, then, are we concerned and involved in its estimation and computation?

Answer: We are concerned and involved because our lives and activities are affected by both the positive energy, associated with mass and the negative energy associated with gravity and the zero energy arising from their interactions.

Example 1.10: What do we know about positive energy?

Answer: What we know about positive energy may be related to what the Standard Model of cosmology defines as the total mass–energy of the universe which is made up of 26.8% of dark matter, 68.3% of dark energy and 4.9% of ordinary matter.

Example 1.11: What is dark energy?

Answer: Dark energy is a hypothetical concept but is the most accepted hypothesis for explaining the accelerating expansion of the cosmic universe. Dark energy is thought to be homogenous, permeates all space and interacts through any of the fundamental forces except gravity. Its density is constant and does not change with the expansion in space.

Two forms of dark energy are recognised a). the cosmological constant and b). quintessence. The cosmological constant is said to be equivalent to vacuum energy, and has negative pressure. It is homogenous and has constant energy density throughout space (10^{-29} g/cm^3). Quintessence, on the other hand, is said to be a scalar and dynamic quantity which varies in time and space (2).

Example 1.12: What is dark matter?

Answer: The consensus of scientific opinion is on things that dark matter is not rather than on what it is (1). It is agreed by all that the cosmic universe is composed of approximately, 68% dark energy, 27% dark matter and 5% normal matter.

Dark matter is not in the form of the stars or planets that we see because observations show that there is too little visible matter to make up the 27% attributed to dark matter.

Dark matter is not in the form of dark clouds of normal matter, made up of particles called baryons, otherwise they would be detectable by their absorption of radiation passing through them.

Dark matter is not anti-matter otherwise we would be able to see the unique gamma rays produced when anti-matter annihilates with matter.

Dark matter is not galaxy sized black holes because the gravitational bending of light by these (lensing) that have been observed are not enough to make up the required 25 % dark matter contribution.

One speculation posits that dark matter may still be baryonic matter tied up in brown dwarfs or in small, dense chunks of heavy elements known as Massive Compact Halo Objects (MACHOs). The more common speculation is that dark matter is not baryonic at all but is made up of more exotic particles such as axions or WIMPS (Weakly Interactive Massive Particles) (2).

Example 1.13: Which, then, is the energy we encounter?

Answer: The energy we encounter is, by implication, the cosmic energy that is associated, according to the Einstein equation, with the 5% of normal or baryonic matter in the cosmic universe. It is manifested in various forms, the most well-known of which are (Classroom-energy.org, 2009, Wikipedia, 2009)

- *Mechanical energy* which includes potential energy or energy stored in a system and kinetic energy or energy due to or arising from motion. Sound is a form of kinetic energy.
- *Radiant or solar energy* which comes from the light and warmth of the sun.
- *Internal energy* which is the sum of all the microscopic forms of energy of a system. It is related to the molecular structure and degree of molecular activity in the system. Internal energy can, also, be considered as the sum of the kinetic and potential energies of the molecules and is made up of sensible energy, latent energy, chemical energy, nuclear energy, thermal energy and interactions of the energies listed above.
- *Thermal energy* which, usually, increases or decreases with the addition or removal of heat to or from an object but is not the same as heat. It is the sum of sensible and latent forms of energy
- *Chemical energy*, which is that stored in the chemical bonds of molecules.
- *Sensible energy* which is the portion of the internal energy of a system associated with kinetic energies of molecular translation, rotation, vibration. It includes, also, electron translation and spin as well as the nuclear spin of the molecules.
- *Latent energy* which is the internal energy associated with the phase of the system
- *Electrical energy* which is associated with the movement of electrons
- *Electromagnetic energy* which is associated with light waves (including radio waves, microwaves, x-rays, infrared waves, etc)
- *Mass or nuclear energy* which is found in the nuclear structure of atoms
- *Energy interactions* which are those types of energy not stored in the system but are only recognised at the system boundary when they cross it either as energy gains or losses during a process. Examples are heat transferred, mass transferred and work

Example 1.14: What natural laws govern the conversion of these various forms of the energy from one to the other?

Answer: The natural laws which govern the conversion of these various forms of energy we encounter, may be summarised as follows:

1. *The law of conservation of energy* (first law of thermodynamics). This states that the total energy of the universe is constant. That is, energy cannot be created or destroyed but can be converted from one form to another, moved from one place to another or be converted into mass.

2. *The entropy law* (a version of the second law of thermodynamics). This states that all things (mass and energy) always tend to a state of disorder. That is, as the disorder increases, energy changes into less usable forms. Put another way, energy conversion can never be 100% efficient.

3. *The law of absolute zero* (a version of the third law of thermodynamics). This states that molecular motion stops at absolute zero (-273 C). In other words, there cannot be a temperature (which measures how fast molecules move) lower than absolute zero.

Example 1.15: What is the law of conservation of energy (the first law of thermodynamics)?

Answer: The global law of the conservation of energy asserts that energy cannot be created or destroyed but can only be transformed.

Most kinds of energy, except gravitational energy, are, in addition, subject to strict local conservation laws that energy can only be exchanged between adjacent regions of space, each of given volumetric energy density.

In more advanced treatments, the law of conservation of energy is regarded as a fundamental principle of physics which follows from the translational symmetry of time. This symmetry asserts that below the cosmic scale, time, for most phenomena, is independent of their locations on the time co-ordinate (this is the physicists' way of saying that time intervals taken at different times are indistinguishable, that is, five seconds interval is still five seconds interval in the morning, in the evening, at night, today, tomorrow, yesterday, etc.). This is possible because energy is the quantity which is the canonical conjugate of time.

This conjugation of time and energy result in an uncertainty principle which makes it impossible to define the exact amount of energy during any definite time interval. It does not negate the conservation law but puts a mathematical limit to the extent which we can, in principle define and measure energy. Please note that this uncertainty principle is different from the Heisenberg uncertainty principle concerned with the position and velocity of elementary particles).

In normal, everyday use, however, the law of conservation of energy, that we know, is that which states that the total inflow of energy into a system must be equal to the total outflow of energy from that system plus any changes of energy contained within the system. That is

$$Energy\ In \pm Energy\ generated\ or\ consumed$$
$$= Energy\ Out \pm Energy\ Accummulated \qquad (1.03)$$

Example 1.16: What are the fundamental and natural methods of conversion of energy?

Answer: The fundamental, natural, conversion methods of energy are

- *Photosynthesis* which is the process used by plants to capture and transform the radiant energy of the sun into the chemical energy they need to grow.
- *Digestion* through which animals capture and burn the chemical energy in food, water and air to fuel their own cell growth (chemical energy) and movement (kinetic energy).

- *Combustion*, a thermo chemical reaction, in which the stored chemical energy in fossil fuels is converted to either heat energy or to kinetic energy of motion or to both.
- *Nuclear reactions* in which the mass energy in atoms is released either in a nuclear fission, nuclear fusion or anti-matter reactions. The energy released is given by $E = mc^2$ where m is the mass and c is the speed of light

Example 1.17: What other methods are used in energy conversions?

Answer: Other energy conversion methods include

- *Kinetic to electrical energy* conversions such as occur in the internal combustion engines or in the gas, water or wind turbine generators.
- *Chemical to electrical energy* conversions such as occur in electrolytic cells, batteries, fuel cells and vice versa in some cases such as in battery recharges and electrolysis.
- *Solar to electrical energy* conversions such as are found in solar cells.
- *Solar to thermal energy* conversions such as are found in solar dryers, solar heaters etc.
- *Sound to thermal energy* conversions such as are found in microwave ovens.

Example 1.18: Are there limits to the scope and direction of these energy conversions or transformations?

Answer; Yes

Example 1.19: What are these limits?

Answer: We have already stated the most important limit, namely, that energy cannot be destroyed or created.

The next most important limit is the limit to the efficiency of conversion or transformation of energy. When energy is to be converted or transformed from heat to work the limit is defined

11

by the second law of thermodynamics which, with respect to engines, is expressed by the limit to efficiency defined by the Carnot's cycle and, in respect of refrigerators, by the limit to the coefficient of performance defined by the Rankine cycle.

In a similar manner, there are limits to chemical energy conversions (overcoming an energy barrier), biological energy conversions (reaction with oxygen), violent natural phenomena (energy transformations in the Earth's interior or atmospheric solar energy triggers).

In terms of what kind of energy can be transformed to what other kind (direction of transformation), the limits are set by entropy. According to this, there must be equal energy spread among all available degrees of freedom. In other words, all energy transformations are permitted on a small scale but certain larger ones are not permitted because it is statistically unlikely that energy or matter will move into more concentrated forms or smaller spaces.

Things begin to get a bit complicated when reversible and irreversible processes, the release of some of the potential energy of the universe stored over time such as in nuclear decay, nuclear fission, and chemical explosions are considered but, as a general rule, each of these phenomena has limits defined by an appropriate scientific observation.

Example 1.20: What are the natural sources of energy?
Answer: Natural sources of energy fall into two broad kinds, renewable sources and non-renewable sources. Some examples of renewable energy sources are:

- *Hydropower* in which the energy of flowing water and ocean waves and tides are used to power mechanical equipment directly or used in hydroelectric plants to generate electricity.

12

- *Biomass* in which natural materials such as wood, agricultural crop waste, fast growing willow and switch grass crops, animal wastes, municipal waste and garbage can be used as renewable sources of energy to generate heat and power or as alternative routes to petrochemicals in making plastics and other products. Typical products are ethanol and methane gas.
- *Passive solar heating* in which solar energy is collected for space heating, hot water heating or for crop drying.
- *Active solar systems* in which photovoltaic cells convert solar energy to electrical energy.
- *Solar thermal electricity* in which highly curved mirrors concentrate solar energy onto pipes containing water. The water is converted to steam which is used to drive turbines to generate electricity.
- *Wind power* uses the energy of winds to turn large blades of windmills to produce direct mechanical power for grinding mills or water pumps or to drive electric generators.
- *Geothermal energy* in which the hot water from below the ground (geysers) are used for space heating or, in some areas, to generate electricity.

Examples of non-renewable sources of energy are:

- Nuclear fuel which occurs in certain rock formations as mostly Uranium-235. Fusion reactions, the type that occurs in the sun, are still in the developmental stages. The practical nuclear reactions are the fission reactions in which the Uranium atom is split to release energy with which high pressure steam is generated and used to drive turbines which generate electricity. Nuclear reactions are clean in the sense that they do not produce carbon dioxide and acid rain but are unclean in the sense that their products are dangerous wastes which remain radioactive for thousands of years and are expensive to store and treat.
- Fossil fuels such as coal, natural gas, crude oil and oil shale are more abundant over all the earth than nuclear fuels. They have to undergo several, and sometimes severe, processes to get them into the forms in which they can be used to generate thermal, mechanical or electrical energy. They are cheaper and more convenient to use than nuclear energy but are associated with the production of polluting substances such as carbon dioxide, acid rain, soot, smoke etc.

13

The choice or choices of which energy to use depend on availability, convenience and ease of use in desired application, reliability of supply, cost and effects on public safety, health and the environment.

Example 1.21: What, in fact, is energy transfer?

Answer: Energy transfer is the transfer of energy from one system to another by either the transfer of matter, since matter and energy are interchangeable, or by means other than the transfer of matter which produce changes in the other system as a result of the work done on it.

Heat energy transfer can occur by both mechanisms as it can occur in physical contact or in the non-physical contact situation of electromagnetic radiation.

These considerations lead to the conclusion, for example, that in an open system, the heat energy transferred, ΔE, is given by

$$\Delta E = W + Q + E_{addition} \qquad (1.04)$$

where W is the net work done, Q is the net heat flow in the system and $E_{addition}$ is the net energy flows of any type carried over the surface of the system or control volume.

Example 1.22: What is the difference between heat energy transfer and heat energy transport?

Answer: Heat energy transport occurs within a phase, and is by means of bulk motion while heat energy transfer takes place across an interface or boundary and as a result of molecular motion.

Example 1.23: What are the fundamental methods of heat

energy transfer or transport?

Answer: There are three fundamental modes of heat transfer or transport namely: conduction, convection and radiation.

Example 1.24: What is heat conduction?

Answer: Heat conduction is the transfer or transport of heat from one part of a body to another in physical contact with it without appreciable displacement of the particles of the body.

It is usually the result of an interchange of kinetic energy between molecules. Typical examples are

    (i)   heat flow from one end of a heated bar to another.
    (ii)   heat flow within a solid food product during heating.

Conduction is, usually, greater in solids, where the atoms are closer to each other, than in liquids (except liquid metals) and gases, where they are farther apart. In fluids, heat conduction occurs, primarily, by elastic impact of the molecules. In metals, heat conduction occurs by free electron diffusion while in insulators, it occurs by phonon vibration.

Metals are the best thermal energy conductors because their kind of bonding (metallic bonds as opposed to covalent or ionic bonds) allows free movement of electrons and formation of crystal structures which, greatly, aid the transfer of thermal energy.
As the density decreases so does conduction because of the large distance between atoms in such a material. Thus liquids and gases are poorer thermal conductors than solids which tend to have greater densities.

Example 1.25: What is heat convection?

15

Answer: Heat convection is the transport of heat from one point to another within a fluid (gas or liquid) by the mixing of one portion of the fluid with another. Often it occurs by the combination of heat conduction and of heat transfer by circulation or movement of hot particles in bulk.

In natural convection, bulk motion is as a result of density differences (buoyancy) which arise from temperature differences, usually, in a gravity field. In forced convection, bulk motion is produced by mechanical means such as a pump etc.

Example 1.26: What is heat or thermal radiation?

Answer: Heat or thermal radiation occurs when energy is transferred by electromagnetic waves from a body at a high temperature to one at a lower temperature which is not in contact with it. No medium is required (vacuum gives the highest transfer). Radiation is significant only at elevated temperatures.

Example 1.27: Why are we interested in heat transfer or transport?

Answer: Our interest in heat transfer or transport arises from our desire and need to understand, to estimate or to control the

- mechanism of heat transfer or transport,
- amount of heat energy transported in any given situation,
- temperature distribution within and/or outside the medium or system in which the heat transfer/transport occurs.
- properties of the medium, or system which affect or is affected by the heat transfer or transport,
- systems which facilitate the transfer and/or transport of heat energy in practical and commercial situations
- strategies which optimise the utilisation of system properties and operating conditions for heat transfer or transport.

References for Chapter One

1. http://en.wikipedia.org/wiki/Energy

2. http://en.wikipedia.org/wiki/Dark energy

3. http://science.nasa.gov/astrophysics/focus-areas/what-is-dark-energy/

4. Classroom-energy.org, 2009, Wikipedia, 2009

CHAPTER TWO
CONDUCTION OF HEAT

Example 2.01: What is heat, or thermal energy, flux?

Answer: Heat, or thermal energy, flux is the flow of heat, or thermal, energy, per unit time per unit area perpendicular to the direction of heat or thermal energy flow. That is

$$q = \frac{Q}{t \ x \ A} \tag{2.01}$$

where Q is the amount of heat energy in time t flowing perpendicular to an area A.

Example 2.02: What is the mathematical description of heat, or thermal energy, flux?

Answer: The flow, by conduction of heat or thermal energy, is a transport process, and like all transport processes, follows the general rule that the rate of transport is proportional to the driving force. That is:

$$Rate \ of \ transport = \frac{Driving \ force}{Resistance}$$

$$= Driving \ force \ x \ Conductance \tag{2.02}$$

where $\qquad Resistance = \dfrac{1}{Conductance}$

and $\qquad Conductance = \dfrac{1}{Resistance}$

Since the heat flux, **q**, is defined as the quantity of heat energy per unit time per unit cross-sectional area normal to the direction of heat energy flow, it follows, by comparison with equation (2.02), that

$$q = \frac{Driving\ force}{Resistance} = \frac{\Delta T}{\dfrac{\Delta x}{k}} = -k\frac{\Delta T}{\Delta x} \qquad (2.03)$$

where k is a proportionality constant representing heat flux per unit temperature difference per unit distance (temperature gradient). Δx is an incremental spatial distance and $\dfrac{\Delta x}{k}$ is, thus, a resistance to conductive heat transfer. In the limit

$$q = -k \lim_{\Delta x \to 0} \left|\frac{\Delta T}{\Delta x}\right| = -k\frac{\partial T}{\partial x} \qquad (2.03a)$$

That is, the rate at which heat crosses an isothermal surface in the direction of falling temperature is proportional to the temperature gradient. This is a statement of Fourier's first law of heat conduction in an isotropic medium. The constant, k, is called the thermal conductivity of the medium.

Example 2.03: What is an isotropic medium?

Answer: An isotropic medium is one whose structure and properties, in the neighbourhood of any point within it, are the same in all directions through the point.

Example 2.04: Derive the governing equations for heat conduction in a solid

Answer: Consider heat conduction in an infinitesimal element of a solid, in Cartesian co-ordinates. With the nomenclature and terms displayed in the sketch, a heat energy balance, in all directions, on the basis of the law of conservation of energy, will show that:

Heat in + Heat generated within element

$$= Heat\ out + Heat\ accumulated\ within\ element \qquad (2.04)$$

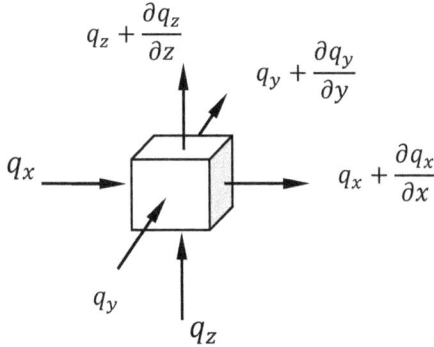

If Q_v is the heat generated per unit volume per unit time, within the element, and q_i is the heat flux, defined in any given direction, i, then:

$q_x . \Delta y . \Delta z + q_y . \Delta z . \Delta x + q_z . \Delta y . \Delta x + Q_v . \Delta x . \Delta y . \Delta z$
Heat entering element + Heat generated within element

$$= \left(q_x + \frac{\partial q_x}{\partial x} . \Delta x\right) . \Delta z. \Delta y + \left(q_y + \frac{\partial q_y}{\partial y} . \Delta y\right) . \Delta z. \Delta x$$

$$+ \left(q_z + \frac{\partial q_z}{\partial z} . \Delta z\right) . \Delta x. \Delta y$$

$$+ \frac{\partial\left(\rho \Delta x. \Delta y. \Delta z. Cp(T - T_0)\right)}{\partial t}$$

Heat leaving element + Heat accumulated within element

Simplifying, with

$$q_x = -k_x \frac{\partial T}{\partial x}, \qquad q_y = -k_y \frac{\partial T}{\partial y}, \qquad q_z = -k_z \frac{\partial T}{\partial z}$$

$$Q_v = \left(\frac{\partial q_x}{\partial x} + \frac{\partial q_y}{\partial y} + \frac{\partial q_z}{\partial z}\right) + \rho Cp \frac{\partial T}{\partial t}$$

$$= -\frac{\partial}{\partial x}\left(k_x \frac{\partial T}{\partial x}\right) - \frac{\partial}{\partial y}\left(k_y \frac{\partial T}{\partial y}\right) - \frac{\partial}{\partial z}\left(k_z \frac{\partial T}{\partial z}\right) + \rho Cp \frac{\partial T}{\partial t} \quad (2.05)$$

This is the general energy equation for heat conduction in a solid and may also be expressed in cylindrical and spherical co-ordinate systems.

Example 2.05: How is the general energy equation for heat conduction affected in an isotropic solid?

Answer: An isotropic solid is one in which all solid properties are the same in all directions. For such a solid, in heat conduction, $k_x = k_y = k_z = k = $ constant and, in the Cartesian co-ordinate system, equation (2.05) reduces to

$$Q_v + k\left(\frac{\partial^2 T}{\partial x^2} + \frac{\partial^2 T}{\partial y^2} + \frac{\partial^2 T}{\partial z^2}\right) = \rho C p \frac{\partial T}{\partial t} \qquad (2.06)$$

Example 2.06: What other forms can the general energy equation for heat conduction in solids take?

Answer: Other forms which the general energy equation, for heat conduction in a solid, can take will depend on

a) whether the solid is isotropic
b) whether heat is absorbed by, or emitted from, the system,
c) whether the conduction is dependent or independent of time

a). For an isotropic solid in heat conduction, $k_x = k_y = k_z = k = $ constant and, in the Cartesian co-ordinate system, we found

$$Q_v + k\left(\frac{\partial^2 T}{\partial x^2} + \frac{\partial^2 T}{\partial y^2} + \frac{\partial^2 T}{\partial z^2}\right) = \rho C p \frac{\partial T}{\partial t} \qquad \text{(eqtn 2.06)}$$

This can be seen to be a special case of the general energy equation (for a solid) when k is constant.

b). If no heat energy is generated by the system, that is, $Q_v = 0$ and k is still constant, we get that

$$\frac{\partial T}{\partial t} = \frac{k}{\rho C p} \left(\frac{\partial^2 T}{\partial x^2} + \frac{\partial^2 T}{\partial y^2} + \frac{\partial^2 T}{\partial z^2} \right) = \alpha \nabla^2 T \qquad (2.07)$$

This is called the Fourier's field equation or Fourier's second law of heat conduction. The thermal diffusivity, α, is given by

$$\alpha = \frac{k}{\rho C p} \qquad (2.08)$$

It is the ratio of conductive to convective heat energy transfer and has the units of length squared per unit time.

c). If energy is generated by the system such that Q_v is not zero but $\frac{\partial T}{\partial t} = 0$, we have the case of steady state heat conduction with internal heat generation. This situation is described by the Poisson's equation:

$$\frac{Q_v}{k} + \left(\frac{\partial^2 T}{\partial x^2} + \frac{\partial^2 T}{\partial y^2} + \frac{\partial^2 T}{\partial z^2} \right) = 0 \qquad (2.09)$$

Equation (2.09) may, also, be expressed, in vector notation, as

$$\frac{Q_v}{k} + \nabla^2 T = 0 \qquad (2.09a)$$

d). If Q_v and $\frac{\partial T}{\partial t}$ are both equal to zero, then there is steady state heat conduction without internal generation of heat. This is the situation described by the Laplace's equation, given below:

$$\frac{\partial^2 T}{\partial x^2} + \frac{\partial^2 T}{\partial y^2} + \frac{\partial^2 T}{\partial z^2} = 0 \qquad (2.10)$$

Example 2.07: How are these equations expressed in other co-

23

ordinate systems such as the Cartesian, cylindrical and spherical co-ordinates.

Answer: Tables 2.1, 2.2 and 2.3 below, summarise the expression of these heat conduction equations in the Cartesian, cylindrical and spherical co-ordinate systems.

Table 2.1: Summary of Heat Conduction Equations in Cartesian Co-ordinates (Isotropic Solids)

1   **Heat Flux Across an Isothermal Surface**

$$q_x = -k_x \frac{\partial T}{\partial x}, \qquad q_y = -k_y \frac{\partial T}{\partial y}, \qquad q_z = -k_z \frac{\partial T}{\partial z}$$

2   **Heat Flux Across any Surface**

$$q \cdot n = -k_n \frac{\partial T}{\partial n}$$

where q = heat flux that flows normal to an isothermal surface and n is the normal to the surface

3   **The General Energy Equation**

$$Q_V + k\left(\frac{\partial^2 T}{\partial x^2} + \frac{\partial^2 T}{\partial y^2} + \frac{\partial^2 T}{\partial z^2}\right) = \rho Cp \frac{\partial T}{\partial t}$$

4   **Poisson's Equation**

$$\frac{Q_V}{k} + \frac{\partial^2 T}{\partial x^2} + \frac{\partial^2 T}{\partial y^2} + \frac{\partial^2 T}{\partial z^2} = 0$$

5   **Laplace's Equation**

$$\frac{\partial^2 T}{\partial x^2} + \frac{\partial^2 T}{\partial y^2} + \frac{\partial^2 T}{\partial z^2} = 0$$

Table 2.2: Summary of Heat Conduction Equations in Cylindrical Coordinates (Isotropic Solids)

1   **Heat Flux Across an Isothermal Surface**

$$q_r = -k_r \frac{\partial T}{\partial r}, \qquad q_\theta = -k_\theta \frac{\partial T}{\partial \theta}, \qquad q_z = -k_z \frac{\partial T}{\partial z}$$

2   **Heat Flux Across any Surface**

$$q.n = -k_n \frac{\partial T}{\partial n}$$

where q = heat flux normal to an isothermal surface and n is the normal to the surface

3   **The General Energy Equation**

$$Q_V + k \left(\frac{\partial^2 T}{\partial r^2} + \frac{1}{r} \frac{\partial T}{\partial r} + \frac{1}{r^2} \frac{\partial^2 T}{\partial \theta^2} + \frac{\partial^2 T}{\partial z^2} \right) = \rho C p \frac{\partial T}{\partial t}$$

4   **Poisson's Equation**

$$\frac{Q_V}{k} + \frac{\partial^2 T}{\partial r^2} + \frac{1}{r} \frac{\partial T}{\partial r} + \frac{1}{r^2} \frac{\partial^2 T}{\partial \theta^2} + \frac{\partial^2 T}{\partial z^2} = 0$$

5   **Laplace's Equation**

$$\frac{\partial^2 T}{\partial r^2} + \frac{1}{r} \frac{\partial T}{\partial r} + \frac{1}{r^2} \frac{\partial^2 T}{\partial \theta^2} + \frac{\partial^2 T}{\partial z^2} = 0$$

Table 2.3: Summary of Heat Conduction Equations in Spherical Co-ordinates (Isotropic Solids)

(i)   **Heat flux across an isothermal surface**

$$q_r = -k_r \frac{\partial T}{\partial r}, \qquad q_\theta = -k_\theta \frac{\partial T}{\partial \theta}, \qquad q_\varphi = -k_\varphi \frac{\partial T}{\partial \varphi}$$

(ii) **Heat Flux across any surface**

$$q \cdot n = -k_n \frac{\partial T}{\partial n}$$

where q = heat flux that flows normal to an isothermal surface and n is the normal to the surface.

(iii) **The General Energy Equation**

$$Q_V + k \left[\frac{1}{r^2} \frac{\partial}{\partial r} \left(r^2 \frac{\partial T}{\partial r} \right) + \frac{1}{r^2 \sin\theta} \frac{\partial}{\partial \theta} \left(\sin\theta \frac{\partial T}{\partial \theta} \right) \right.$$
$$\left. + \frac{1}{r^2 \sin^2\theta} \frac{\partial^2 T}{\partial \varphi^2} \right] = \rho C p \frac{\partial T}{\partial t}$$

(iv) **Poisson's Equation**

$$\frac{Q_V}{k} + \frac{1}{r^2} \frac{\partial}{\partial r} \left(r^2 \frac{\partial T}{\partial r} \right) + \frac{1}{r^2 \sin\theta} \frac{\partial}{\partial \theta} \left(\sin\theta \frac{\partial T}{\partial \theta} \right) + \frac{1}{r^2 \sin^2\theta} \frac{\partial^2 T}{\partial \varphi^2} = 0$$

(v) **Laplace's Equation**

$$\frac{1}{r^2} \frac{\partial}{\partial r} \left(r^2 \frac{\partial T}{\partial r} \right) + \frac{1}{r^2 \sin\theta} \frac{\partial}{\partial \theta} \left(\sin\theta \frac{\partial T}{\partial \theta} \right) + \frac{1}{r^2 \sin^2\theta} \frac{\partial^2 T}{\partial \varphi^2} = 0$$

Example 2.08: What are these equations used for?

Answer: They are used to solve problems in pure conduction or those dominated by pure conduction under the prevailing initial and boundary conditions. The choice of the co-ordinate system to use depends on the physical characteristics of the problem.

If, for example, the object of the heat conduction is rectangular, whether in two or three dimensions, the Cartesian coordinate system equations are recommended. For cylindrical objects such as pipes or rods, use the polar or cylindrical co-ordinate system equations while for drops and spherical objects, the spherical co-ordinate system equations are to be preferred.

Example 2.09: In what areas of heat conduction are

(a)   the Fourier's field equation
(b)   the Poisson equation
(c)   the Laplace's equation.

most usefully applied?

Answer: The Fourier's field equation is best applied to the heating and cooling of solids which involve a time element and or transients in heat transfer.

The Poisson's equation applies to steady state heating with internal heat generation such as occur with electric heating elements, nuclear fuel rods, etc., which may or may not be embedded in an enclosure (solid, liquid or gas).

The Laplace's equation is applied to general steady heat conduction in slabs, plates and cylinders such as pipes, furnace walls, etc. in which no internal heat energy is generated.

Example 2.10: What are the common initial and boundary conditions encountered in the use of these equations?

Answer: The common initial conditions, applicable to the time dependent, general energy and Fourier field, equations, are:

(i). $t = 0$, $T = T_0$, an initial temperature, everywhere constant (2.11)
(ii) $t = 0$, $T = T(x, y, z)$, an initial temperature distribution (2.12)

The common boundary conditions, that can be applied to all of the equations, depending on the problem, are:-

(i) isothermal boundaries, (T is constant at these boundaries)   (2.13)

(ii) insulated boundary ($q = 0$; no heat flux across the boundary)

$$(2.14)$$

(iii) surface at constant heat flux

$$q = -k_x \frac{\partial T}{\partial x}\bigg|_{x=0} = h(T_h - T) \qquad (2.15)$$

where h = heat transfer coefficient.

Note that initial and boundary conditions may or may not apply, simultaneously, to any one problem.

Example 2.11: Are there any other kinds of solids, other than isotropic solids in which heat conduction can take place?

Answer: Yes. These are solids which are not isotropic. In scientific nomenclature they are known as anisotropic non-crystalline solids and anisotropic crystalline solids. They are solids in which the non-uniformity of physical and other properties are mainly due to their being either non-crystalline or crystalline.

Example 2.12: How is their thermal conductivity affected by the nature of these non-isotropic solids?

Answer: a). Anisotropic Non-Crystalline Solids

In anisotropic, non-crystalline solids, the thermal conductivity, k, is different for each spatial direction. That is:

i	$k_x \neq k_y \neq k_z$	in Cartesian co-ordinates
ii	$k_r \neq k_\theta \neq k_z$	in cylindrical co-ordinates
iii	$k_r \neq k_\theta \neq k_\varphi$	in spherical co-ordinates

Thus, for this type of solid, in Cartesian co-ordinates, the Fourier's field equation, for example, becomes

$$k_x \frac{\partial^2 T}{\partial x^2} + k_y \frac{\partial^2 T}{\partial y^2} + k_z \frac{\partial^2 T}{\partial z^2} = \rho C p \frac{\partial T}{\partial t} \qquad (2.16)$$

b). Anisotropic Crystalline Solids

In anisotropic, crystalline solids, the uniqueness of the thermal conductivity is no longer in each spatial, x, y, z, direction, as in an anisotropic non-crystalline solids, but is found in each crystal orientation. Consequently, the heat flux **q**, is not equal, say in the x-direction, to $q_x = -k_x \frac{\partial T}{\partial x}$ but is computed along crystal directions. Thus

$$q_x = -k_{11} \frac{\partial T}{\partial x} - k_{12} \frac{\partial T}{\partial y} - k_{13} \frac{\partial T}{\partial z} \qquad (2.17a)$$

$$q_y = -k_{21} \frac{\partial T}{\partial x} - k_{22} \frac{\partial T}{\partial y} - k_{23} \frac{\partial T}{\partial z} \qquad (2.17b)$$

$$q_z = -k_{31} \frac{\partial T}{\partial x} - k_{32} \frac{\partial T}{\partial y} - k_{33} \frac{\partial T}{\partial z} \qquad (2.17c)$$

where k_{ij} is the thermal conductivity along the crystal structure orientation, *ij*. In such a case, the Fourier field equation, for example, becomes quite complex and can only be handled in its basic form as:

$$\rho C p \frac{\partial T}{\partial t} = \frac{\partial q_x}{\partial x} + \frac{\partial q_y}{\partial y} + \frac{\partial q_z}{\partial z} \qquad (2.18)$$

Very often, however, empirical correlations, which assume isotropy, are developed for both anisotropic non-crystalline and anisotropic crystalline systems in order to exploit the greater simplicity offered by the isotropic heat conduction equations.

Example 2.13: What are the units of the thermal conductivity?

Answer: The units for thermal conductivity may be derived from the rearrangement of equation (2.03) such that

$$k = \frac{Q}{A \cdot t \cdot \frac{\Delta T}{L}} \qquad (2.19)$$

which is seen to have the units of

$$\frac{Joules}{Area \; x \; Time \; x \; \frac{Temperature \; difference}{Thickness}}$$

or $J/m^2 \cdot s$ per (K/m), that is W/m^2 per (deg K per m) or heat flux per unit temperature gradient. A useful and convenient simplification is that all this reduce to $W/m.K$ which may, again, be interpreted as the heat flow per unit time per unit length per unit temperature difference.

Example 2.14: Does the thermal conductivity of a substance have any other implication with respect to the thermal properties of that substance?

Answer: Yes. In addition to being a convenient, mathematical, proportionality, constant in the heat conduction equations, thermal conductivity is also a material property by means of which we can get an idea of how good, or how poor, any particular material is as a thermal conductor or insulator.

It follows that thermal conductivity, as a material property, will, also, depend on the phase of the material, its temperature, density and molecular bonding.

Example 2.15: What are thermal insulators?

Answer: Thermal insulators are, generally, solids, but may be liquids or gases, which reduce the flow of heat by limiting conduction, convection or both. Their thermal conductivities depend, of course, on whether they are isotropic or non-isotropic. The effectiveness of a thermal insulator is measured by its resistance value which is the inverse of its thermal conductivity divided by its thickness. Thermal insulators are different from radiant barriers or radiation shields which reduce the flow of heat from radiation sources by reflecting radiation. Good insulators are not, necessarily, good radiation barriers.

Example 2.16: How would you judge from the value of the thermal conductivities of two different solids which of them would be an insulator?

Answer: Since $q = k\frac{\Delta T}{\Delta x}$, then $k = \frac{q}{\Delta T/\Delta x}$. Thus, for the same amount of heat flux, the solid with a higher thermal conductivity would show a lower temperature gradient and thus less resistance to heat transfer. That is, it is a good conductor of heat. Put another way, this solid would conduct more heat for the same temperature gradient than one with a smaller k.

Similarly, the solid with the smaller k would show larger temperature gradients and thus higher resistance to heat transfer. It would then be a better heat insulator than the solid with a higher thermal conductivity.

References For Chapter Two

1   Bird, R.B; Stewart, W.E; Lightfoot, E.N; *Transport Phenomena*, Chapters 8 – 14; Wiley Int'l Edition, NY., USA 1960.

2   Carslaw, H.S and Jaeger, J.C; *Conduction of Heat in Solids*, Clarendon Press, Oxford, UK, 1959.

CHAPTER THREE
THERMAL CONDUCTIVITY

Example 3.01: We found that there are three types of solids we have to deal with in heat conduction, namely, isotropic, non-isotropic crystalline and non-isotropic non-crystalline solids. How do we estimate their thermal conductivities?

Answer: For isotropic solids, for which the thermal conductivity is the same in every point and direction in the solid, the problem is the straightforward one of determining one value of thermal conductivity, the various methods available which are listed below.

For non-isotropic, non-crystalline solids, however, Carslaw and Jaeger (1959), suggested that, by setting

$$x^1 = x\sqrt{\frac{k}{k_x}}, \quad y^1 = y\sqrt{\frac{k}{k_y}}, \quad z^1 = z\sqrt{\frac{k}{k_z}} \qquad (3.01)$$

where k is a transformed thermal conductivity, the Fourier field equation, for anisotropic, non-crystalline solid, can be transformed to that of the isotropic solid with a new non-directional thermal conductivity k but with new spatial co-ordinates x_1, y_1 and z_1. This non-directional, transformed, thermal conductivity may, then, be determined as for isotropic thermal conductivity.

For non-isotropic, crystalline solids, the heat flux can be solved for in a matrix equation of the type

$$\begin{pmatrix} -k_{11} & -k_{12} & -k_{13} \\ -k_{21} & -k_{22} & -k_{23} \\ -k_{31} & -k_{32} & -k_{33} \end{pmatrix} \begin{pmatrix} \dfrac{\partial T}{\partial x} \\ \dfrac{\partial T}{\partial y} \\ \dfrac{\partial T}{\partial z} \end{pmatrix} = \begin{pmatrix} q_x \\ q_y \\ q_z \end{pmatrix} \qquad (3.02)$$

Knowledge of the heat flux and temperature gradient in each crystal direction make it possible to evaluate this matrix and hence estimate the respective thermal conductivities in each crystal orientation.

Example 3.02: What are the best values of thermal conductivity to use?

Answer: The best values of the thermal conductivity to use are those obtained from experimental measurements because they represent actual physical conditions. Experimental values apply, however, only to the specific conditions in which they were determined and may not be applicable to other situations. Sometimes there are no experimentally determined values of thermal conductivity available for the material.

In both cases, mathematical correlations have to be resorted to in order to estimate either the value or the form of k. These correlations have been presented in several text and reference books such as those by Bird *et al* (1960), Perry & Green (1984), Smith (1959) and Heldman (1975).

Example 3.03: How are the thermal conductivities of gases estimated in the absence of reliable experimental data?

Answer: For pure gases, kinetic theory analysis shows k to be proportional to $\sqrt{T}$ and independent of pressure. For a monatomic gas, Bird *et al* (1960) give

$$k = \frac{1}{d^2} \sqrt{\frac{k^3 T}{\pi^3 m}} \tag{3.03}$$

where d = molecule diameter, k = Boltzman constant, m = mass of molecule.

This expression gives only approximate values for k.

The Chapman-Enskog formula, for monatomic gas, at low density (less than 10 atm.), is

$$k = \frac{0.0829}{\sigma \Omega_k} \sqrt{\frac{T}{M}} \tag{3.04}$$

where k = W/m. K, σ = collision diameter, A°, Ω_k = Lennard Jones collision integral (from Tables), M = molecular weight.

The value of k obtained, with this expression, is fairly accurate.

For polyatomic gases, the Eucken formula for a non-polar gas, at low density, is:

$$k = \mu C_V + \left(\frac{9R}{4M}\right) = \mu \left(Cp + \frac{5R}{4M}\right) = \mu \left(Cp + \frac{2.48}{M}\right) \tag{3.05}$$

where R is gas constant, M is molecular weight, μ is viscosity, Cp is heat capacity at constant pressure, Cv is heat capacity at constant volume.

Average errors are: 5% for a non-polar, linear gas, 8% for a non-polar, non-linear gas and 13% for a polar, non-linear gas. Maximum error lies between 25 - 30%.

Bromley's correlation is empirical and more accurate. For monatomic gases, it is

$$\frac{Mk}{\mu C_V} = 2.4 + 0.016\sqrt{M} = 2.5 \qquad (3.06)$$

For non-polar linear molecules, it is

$$\frac{Mk}{\mu} = 1.32 C_V + 3.40 - \frac{0.70}{T_R} \qquad (3.07)$$

For non-polar or polar non-linear molecules, it is

$$\frac{Mk}{\mu} = 1.30 C_V + 3.60 - 0.3 C_{IR} - \frac{0.69}{T_R} - 3\alpha \qquad (3.08)$$

where k is in calories/s.cm, μ is in poise, g/cm.s. Cv, C_{IR}, α are in calories/gmole C, C_{IR} is the internal rotational heat capacity contribution (obtained from Tables).

$$\alpha = 3.0 \frac{\rho_b}{M} \left(\frac{\lambda_b}{T_b} - 8.75 - R \ln T_h \right) \qquad (3.09)$$

ρ = density g/cc, λ_b = latent heat cal/g. mole, T_b = boiling point in °K, $\qquad$ R = 1.987 cal/gmole K

This equation is not to be used for hydrogen halides. Average error is 4% for non-polar linear molecules, 3.5% for non-polar non-linear and 7.5% for polar non-linear molecules. Maximum error is 10%.

For gas mixtures, the Lindsay - Bromley correlation is:

$$k_x = \sum_{i=1}^{n} \frac{x_i k_i}{\sum_{j=1}^{n} x_i \varphi_{ij}} \qquad (3.10)$$

where

$$\varphi_{ij} = \frac{1}{\sqrt{8}}\left(1+\frac{M_i}{M_j}\right)^{-\frac{1}{2}}\left[1+\left(\frac{\mu_i}{\mu_j}\right)^{\frac{1}{2}}\left(\frac{M_j}{M_i}\right)^{\frac{1}{4}}\right]^2 \qquad (3.10a)$$

Average error is 4% for mixtures of non-polar polyatomic gases such as CH_4, O_2, N_2, C_2H_2 and CO.

A less complicated yet fairly accurate equation is given by:

$$k_x = \frac{\sum x_i k_i M_i^{1/3}}{\sum x_i M_i^{1/3}} \qquad (3.11)$$

Average error is 2.7%. with maximum error of 9.5% for binary mixtures at low pressure.

Example 3.04: How are the thermal conductivities of liquids estimated in the absence of reliable experimental data?

Answer: For liquids the modified Bridgman equation gives

$$k = 2.80\left(\frac{N}{V}\right)^{2/3} k_b V_s \qquad (3.12)$$

where V/N = volume per molecule; k_b = Boltzman constant and V_s = velocity of sound. This equation is fairly accurate even for polyatomic liquids in the neighbourhood of the critical density.

The Vergaftik's modification (Palmer's expression) is:

$$k = \frac{1.034 Cp.\rho^{4/3}}{\alpha M^{1/3}}, \qquad BTU/h.ft.^o F \qquad (3.13)$$

Cp is in Btu/lb.F and ρ is in g/cc. α is an abnormality factor equal to $\lambda/(21T_b)$. α is evaluated at 30 C for most liquids or at $T_c/2$ for low boiling liquids. α is rounded off to 1 if the calculated value is less than 1. Above 30°C the value of k is obtained by interpolation with $\alpha = 1$ at T_c where T_c = critical temperature, deg K. Average error is 8.7% with maximum error of 31.6%

Example 3.05: How are the thermal conductivities of liquid mixtures estimated in the absence of reliable experimental data?

Answer: For petroleum fractions and oil mixtures, Smith (1959) states that $k = 0.079$ Btu/h ft °F at 30 C + 3% error, while Cragoe states that

$$k = \frac{0.0677}{S.G.}[1 - 0.0003(t - 32)] \qquad (3.14)$$

Here k is in Btu/h.ft.°F, the specific gravity, S.G, is that at 60°F/60°F and t is in °F. This expression is valid for $32 < t < 392$ °F and $0.78 < S.G < 0.95$. Average error is 12% with maximum error of 39%.

Example 3.06: How are the thermal conductivities of miscible liquid mixtures estimated in the absence of reliable experimental data?

Answer: These are estimated from the Barratt-Nettleton function given by (International Critical Tables Vol. 5, p 227)

$$k_m sinh(100b) = \sum k_i sinh(w_i b) \qquad (3.15)$$

where b is a constant specific to the constituents and w_i is weight percent of component i.

Example 3.07: How are the thermal conductivities of binary non-polar liquid mixtures estimated in the absence of reliable experimental data?

Answer: These are estimated from the Filippo and Novoselova equation (Heldman, 1975), given by

$$k_m = k_1 w_1 + k_2 w_2 - 0.72(k_2 - k_1)w_1 w_2 \qquad (3.16)$$

w_2 refers to the component for which $k_2 > k_1$

Example 3.08: How are the thermal conductivities of solid-liquid suspensions estimated in the absence of reliable experimental data?

Answer: These are estimated from the Tareef's equation given by

$$k_m = k_c \frac{2k_c + k_d - 2\varphi_d(k_c - k_d)}{2k_c + k_d + \varphi_d(k_c - k_d)} \qquad (3.17)$$

where φ is phase volume fraction, subscripts c and d refer to the continuous and discontinuous phases. This equation applies, also and with fair accuracy, to liquid-liquid emulsions.

Example 3.09: How are the thermal conductivities of solids estimated in the absence of reliable experimental data?

Answer: These are estimated

i. For Pure Metals, by the Weidman, Franz and Lorenz equation, which is stated as

$$\frac{k}{k_e T} = L = constant \qquad (3.18)$$

k is thermal conductivity at temperature T,°K, k_e is electrical conductivity at temperature T,°K, L is the Lorenz Number = 22-29 x 10^{-9} (Volt)^2(deg K)$^{-2}$

ii. For Pure or Alloyed Metals, Solid or Liquid, the applicable expression is

$$k = 2.61x10^{-8}\frac{T}{\rho_e} - \frac{2x10^{-17}T^2}{Cp.\rho.\rho_e^2} + \frac{97Cp.\rho^2}{M.T} \qquad (3.19)$$

where k is in W/cm°C , T is in °K, ρ_e is electrical resistivity, ohm -

cm., ρ is density in g/cc, Cp is in cal/g°C and M is the average atomic weight, g/g atom or g/gmole. Maximum error is between 5 and 10 per cent.

iii.   For Woods, when heat flow is normal to the grain (Green, 1984), the expression is

$$k = \rho(0.1159 + 0.00233M) + 0.01375, BTU/h.ft.^o F \qquad (3.20)$$

where ρ, (g/cc) is the weight when dry over volume when green. M is the moisture content of the wood and has to be less than 40% by weight. When M > 40%, a different expression is used, given by

$$k = \rho(0.1159 + 0.00316M) + 0.01375, BTU/h.ft.^o F \qquad (3.21)$$

iv   For Porous Media (solid + liquid or gas), the Russell equation is used and is given as

$$\frac{k_m}{k_{cont}} = \frac{v\rho^{2/3} + 1 - \rho^{2/3}}{v(\rho^{2/3} - \rho) + 1 - \rho^{2/3} + \rho} \qquad (3.22)$$

where

$$\rho = \frac{\rho_{solid} - \rho_m}{\rho_{solid} - \rho_{gas}} \qquad (3.22a)$$

and m represents composite phase, *cont* represents continuous phase and

$$v = \frac{k_{porosities}}{k_{cont}} \qquad (3.22b)$$

The use of this equation assumes that the thermal conductivities of the continuous and discontinuous phases are known.

Example 3.10: What is the effect of temperature on thermal conductivity?

Answer: At any temperature T_2

$$k_{T_2} = \frac{\beta_{T_1}}{\alpha_{T_2}} \rho_{T_2}^{4/3} \qquad (3.23)$$

where β_{T_1} is a constant determined from known k_l at temperature T_1.

A convenient way of obtaining k as a function of temperature is from a plot of k versus a reduced temperature defined as $\theta = \frac{T-T_f}{T_C-T_f}$ which, usually, results in a linear curve. Note that T_f is a reference temperature such as T_1.

Example 3.11: What is the effect of pressure on thermal conductivity?

Answer: The effect of pressure on thermal conductivity is negligible at pressures less than 34 atmospheres. Above 34 atmospheres, use Lenoir's generalised correlation

$$\frac{k_{p_2}}{k_{p_1}} = \frac{\varepsilon_2}{\varepsilon_1} \qquad (3.24)$$

where $P_2 > P_1$ and ε is the conductivity factor obtained from tables in which ε is plotted against T_R and P_R. T_R and P_R are reduced temperature and pressure.

Example 3.12: Aren't there common general and empirical correlations for estimating thermal conductivity?

Answer: Yes. There are.

a. For solids in general, the correlation is

$$k = k_o(1 + \beta T) \qquad (3.25)$$

β is negative for good conductors except aluminium and positive for good insulators.

b. For Food Materials, the correlations are

    i.   The Reidel equation, useful for fruit juices and sugar solutions only.

$$k = (307 + 0.645T - 0.00104T^2)(0.46 + 0.054X)x10^{-3} \qquad (3.26)$$

where X = % water and T is in °F

    ii.   The Kopelman equations

    a). For Isotropic Two Component medium

$$k = k_L \left[\frac{(1 - M^2)}{1 - M^2(1 - M)} \right] \qquad (3.27)$$

M is the volume fraction of solids or discontinuous phase in the product and $k_L > k$. When $k > k_L$,

$$k = k_L \left[\frac{(1 - \varphi)}{1 - \varphi(1 - M)} \right] \qquad (3.28)$$

where

$$\varphi = M^2 \left(1 - \frac{k_S}{k_L}\right) \qquad (3.28a)$$

and k_L is for the continuous phase while k_s is for the discrete phase.

b). For Anisotropic Two Component Media such as Fibrous Materials

For heat conduction parallel to the fibres

$$k_{11} = k_L \left[1 - N^2 \left(1 - \frac{k_S}{k_L}\right)^2\right] \qquad (3.29)$$

For heat conduction perpendicular to the fibres

$$k_{\perp} = k_L \left[\frac{(1 - \varphi)}{1 - (1 - N)}\right] \qquad (3.30)$$

where $\varphi = N\left(1 - \frac{k_S}{k_L}\right)$ and N = volume fraction of solids.

c). For Disperse Phase, No Fibres

Maxwell equation

$$k = k_L \left[\frac{1 - b\left(1 - a\frac{k_S}{k_L}\right)}{1 + b(a - 1)}\right] \qquad (3.31)$$

where

$$a = \frac{3k_L}{2k_L + k_S} \qquad (3.31a)$$

$$b = \frac{X_S}{X_L + X_S} \qquad (3.31b)$$

43

and X_s, k_s and X_L, k_L are water content and thermal conductivity of the solid and liquid phases, respectively.

Example 3.13: The thermal conductivity of asbestos between 300K and 600K is given by $k = 0.138 (1 + 1.95 \times 10^{-4} T)$, W/m.K. Determine the mean thermal conductivity of asbestos over this temperature range.

Answer: Given that $k = 0.138 (1 + 1.95 \times 10^{-4}T)$ between 300 K and 600 K, the mean thermal conductivity can be obtained from the general averaging procedure of mathematics. Thus, the mean thermal conductivity, k_m, is given by

$$k_m = \frac{\int_{T_1}^{T_2} k dT}{\int_{T_1}^{T_2} dT} = \frac{\int_{T_1}^{T_2} 0.138(1 + 1.95 \times 10^{-4}T)}{\int_{T_1}^{T_2} dT}$$

That is

$$k_m = \frac{0.138 \left| T + 1.95 \times 10^{-4} \frac{T^2}{2} \right|_{300}^{600}}{|T|_{300}^{600}}$$

$$= \frac{0.138(600 - 300) + 1.95 \times 10^{-4} (600^2 - 300^2)/2}{(600 - 300)}$$

or

$$k_m = 0.138 \times 1.08775 = 0.150 \ W/m.K \qquad Ans$$

Alternatively, k can be evaluated at either the mean temperature of 450 K or at each of these temperatures and the mean of the values taken.

Thus, at 300 K, $k_{300} = 0.138(1 + 195 \times 10^{-4} \times 300) = 0.146$ W/m.K

At 600K, $k_{600} = 0.138(1 + 1.95 \times 10^{-4} \times 600) = 0.154$ W/m.K

$$Mean\ k = k_m = \frac{k_{300} + k_{600}}{2} = \frac{0.146 + 0.154}{2}$$
$$= 0.150\ W/m.K \qquad Ans$$

Example 3.14:   The Lindsay-Bromley correlation for the thermal conductivity of gaseous mixtures is given by

$$k_x = \sum_{i=1}^{n} \frac{x_i k_i}{\sum_{i=1}^{n} x_i \varphi_{ij}} \qquad (3.32)$$

where

$$\varphi_{ij} = \frac{1}{\sqrt{8}} \left(1 + \frac{M_i}{M_j}\right)^{-1/2} \left[1 + \left(\frac{\mu_i}{\mu_j}\right)^{1/2} \left(\frac{M_j}{M_i}\right)^{1/4}\right]^2 \qquad (3.33)$$

Alternatively

$$k_x = \frac{\sum x_i k_i M_i^{1/3}}{\sum x_i M_i^{1/3}} \qquad (3.34)$$

Estimate the thermal conductivity of a flue gas mixture consisting of

Nitrogen	73.290 %
Carbon Dioxide	11.368 %
Water	14.457 %
Oxygen	0.886 %

using these correlations and at 600K.

Answer: The molecular weight and viscosities of these compounds, at 600 K, are (Welty, 1978),

Compound	M_i	μ, Ns/m^2	k, W/mK
N_2	28	29.127x10^{-6}	4.5549 x10^{-2}

CO_2	44	26.827 x 10^{-6}	4.3097 x10^{-2}
H_2O (steam)	18	206.4 x 10^{-6}	4.2161 x10^{-2}
O_2	32	33.93 x 10^{-6}	4.8364 x 10^{-2}

Using equation (3.33) of the Lindsay-Bromley correlation:

$$\varphi_{N_2-CO_2} = \frac{1}{\sqrt{8}}\left(1+\frac{28}{44}\right)^{-1/2}\left[1+\left(\frac{29.127}{26.827}\right)^{1/2}\left(\frac{44}{28}\right)^{1/4}\right]^2$$

$$= 0.3536 \ x \ 1.7817 \ x \ [1 + 1.04199 \ x \ 1.1196]2 \ = 1.2975$$

$$\varphi_{N_2-H_2O} = \frac{1}{\sqrt{8}}\left(1+\frac{28}{18}\right)^{-1/2}\left[1+\left(\frac{29.127}{206.4}\right)^{1/2}\left(\frac{18}{28}\right)^{1/4}\right]^2$$

$$= 0.3536 \ x \ 0.6255 \ x \ [1 + 0.3757 \ x \ 0.8954]2 \ = 0.3950$$

$$\varphi_{N_2-O_2} = \frac{1}{\sqrt{8}}\left(1+\frac{28}{32}\right)^{-1/2}\left[1+\left(\frac{29.127}{33.93}\right)^{1/2}\left(\frac{32}{28}\right)^{1/4}\right]^2$$

$$= 0.3536 \ x \ 0.7303 \ x \ [1 + 0.9265 \ x \ 1.0340]2 \ = 0.9900$$

$$\varphi_{CO_2-H_2O} = \frac{1}{\sqrt{8}}\left(1+\frac{44}{18}\right)^{-1/2}\left[1+\left(\frac{26.827}{206.4}\right)^{1/2}\left(\frac{18}{44}\right)^{1/4}\right]^2$$

$$= 0.3536 \ x \ 0.5388 \ x \ [1 + 0.3605 \ x \ 0.7998]2 = 0.3162$$

$$\varphi_{CO_2-O_2} = \frac{1}{\sqrt{8}}\left(1+\frac{44}{32}\right)^{-1/2}\left[1+\left(\frac{26.827}{33.93}\right)^{1/2}\left(\frac{32}{44}\right)^{1/4}\right]^2$$

$$= 0.3536 \ x \ 0.6489 \ x \ [1 + 0.8892 \ x \ 0.9235]2 \ = 0.7610$$

$$\varphi_{H_2O-O_2} = \frac{1}{\sqrt{8}}\left(1+\frac{18}{32}\right)^{-1/2}\left[1+\left(\frac{206.4}{33.93}\right)^{1/2}\left(\frac{32}{18}\right)^{1/4}\right]^2$$

$$= 0.3536 \ x \ 0.8 \ x \ [1 + 2.4664 \ x \ 1.1547]2 \ = 4.1885$$

$$\varphi_{O_2-H_2O} = \frac{1}{\sqrt{8}}\left(1+\frac{32}{18}\right)^{-1/2}\left[1+\left(\frac{33.93}{206.4}\right)^{1/2}\left(\frac{18}{32}\right)^{1/4}\right]^2$$

$$= 0.3536 \times 0.6 \times [1 + 0.4054 \times 0.8660]2 = 0.3873$$

$$\varphi_{O_2-CO_2} = \frac{1}{\sqrt{8}}\left(1 + \frac{32}{44}\right)^{-1/2}\left[1 + \left(\frac{33.93}{26.827}\right)^{1/2}\left(\frac{44}{32}\right)^{1/4}\right]^2$$

$$= 0.3536 \times 0.7609 \times [1 + 1.1246 \times 1.0828]2 = 1.3234$$

$$\varphi_{O_2-N_2} = \frac{1}{\sqrt{8}}\left(1 + \frac{32}{28}\right)^{-1/2}\left[1 + \left(\frac{33.93}{29.127}\right)^{1/2}\left(\frac{28}{32}\right)^{1/4}\right]^2$$

$$= 0.3536 \times 0.6831 \times [1 + 1.0793 \times 0.9672]2 = 1.0091$$

$$\varphi_{CO_2-N_2} = \frac{1}{\sqrt{8}}\left(1 + \frac{44}{28}\right)^{-1/2}\left[1 + \left(\frac{26.827}{29.127}\right)^{1/2}\left(\frac{28}{44}\right)^{1/4}\right]^2$$

$$= 0.3536 \times 0.6236 \times [1 + 0.9597 \times 0.8932]2 = 0.7606$$

$$\varphi_{H_2O-CO_2} = \frac{1}{\sqrt{8}}\left(1 + \frac{18}{44}\right)^{-1/2}\left[1 + \left(\frac{206.4}{26.827}\right)^{1/2}\left(\frac{44}{18}\right)^{1/4}\right]^2$$

$$= 0.3536 \times 0.8424 \times [1 + 2.7738 \times 1.2504]2 = 5.9474$$

$$\varphi_{H_2O-N_2} = \frac{1}{\sqrt{8}}\left(1 + \frac{18}{28}\right)^{-1/2}\left[1 + \left(\frac{206.4}{29.127}\right)^{1/2}\left(\frac{28}{18}\right)^{1/4}\right]^2$$

$$= 0.3536 \times 0.7802 \times [1 + 2.6620 \times 1.118]2 = 4.3545$$

These results may be summarised as follows

	i	j = N_2	CO_2	H_2O	O_2
1.	N_2	1.0000	1.2975	0.3950	0.9900
2.	CO	0.7606	1.0000	0.3162	0.7610
3.	H_2O	4.3545	5.9474	1.0000	4.1885
4.	O_2	1.0091	1.3234	0.3873	1.0000

Hence, substituting in equation (3.32)

$$k_x = \sum_{i=1}^{n} \frac{x_i k_i}{\sum_{i=1}^{n} x_i \varphi_{ij}}$$

47

k_x

$$= \frac{0.7329 \; x \; 0.045549}{0.7329 \; x \; 1.00 + 1.2975 \; x \; 0.11368 + 0.3950 \; x \; 0.1445 + 0.9900 \; x \; 0.00886}$$

$$+ \frac{0.11368 \; x \; 0.043097}{0.7329 \; x \; 0.7606 + 0.11368 \; x \; 1.00 + 0.14457 \; x \; 0.3162 + 0.00886 \; x \; 0.7610}$$

$$+ \frac{0.14457 \; x \; 0.042161}{0.7329 \; x \; 4.3545 + 0.11368 \; x \; 5.9474 + 0.14457 \; x \; 1.00 + 0.00886 \; x \; 4.1885}$$

$$+ \frac{0.00886 \; x \; 0.04836}{0.7329 \; x \; 1.0091 + 0.11368 \; x \; 1.3234 + 0.14457 \; x \; 0.3873 + 0.00886 \; x \; 1.00}$$

That is

$$k_x = 0.035278 + 0.006771 + 0.001505 + 0.000449$$

$$= 0.044003 \; W/m.K \quad Ans$$

Using the alternative correlation, equation (3.34)

$$k_x = \frac{\sum x_i k_i M_i^{1/3}}{\sum x_i M_i^{1/3}}$$

Then

k_x

$$= \frac{0.7329x0.045549x(28)^{1/3} + 0.11368x0.043097x(44)^{1/3}}{0.7329x(28)^{1/3} + 0.11368x(44)^{1/3} + 0.14457x(18)^{1/3} + 0.00886x(32)^{1/3}}$$

$$+ \frac{0.14457x0.042161x(18)^{1/3} + 0.00886x0.048364x(32)^{1/3}}{0.7329x(28)^{1/3} + 0.11368x(44)^{1/3} + 0.14457x(18)^{1/3} + 0.0088 \; x(32)^{1/3}}$$

That is

$$k_x = \frac{0.101370 + 0.017296 + 0.015974 + 0.001360}{2.225516 + 0.401330 + 0.378881 + 0.028129}$$

$$= 0.044827, \; W/m.K \quad Ans.$$

48

This is only 1.87% different from the more complicated computation.

Example 3.15: The Tareef equation for the thermal conductivity of solid liquid suspensions is given as

$$k_m = k_c \frac{2k_c + k_d - 2\varphi_d(k_c - k_d)}{2k_c + k_d + \varphi_d(k_c - k_d)}$$

where φ is phase volume fraction. Subscripts c and d denote the continuous and disperse phases. Determine the thermal conductivity of a 50% (by volume) $CaSO_4$ - water suspension at 27 C

Answer: From Welty (1978), the thermal conductivity of water at 27 C is 0.611 W/m.K. From Perry *et al* (1963), the thermal conductivity of $CaSO_4$ is 0.22 Btu/h.ft.F. This value can be converted to SI units by multiplying by 1.7307 W/m.K/Btu/h.ft.F. The converted value is 0.3803 W/m.K

Let water be the continuous phase ie k_c = 0.611 W/m.K and $CaSO_4$ be the dispersed phase ie k_d = 0.381 W/m.K. Then

$$k_m = 0.611 \left(\frac{2x0.611 + 0.381 - 2x0.5\,(0.611 - 0.381)}{2x0.611 + 0.381 + 0.5\,(0.611 - 0.381)} \right)$$

$$= 0.488\,W/m.K, \qquad Ans$$

Example 3.16: The thermal conductivity of pure metals is given by the Weidman, Franz, Lorenz equation as:

$$\frac{k}{k_e T} = 29\,x\,10^{-9}, (Volts)^2/(\deg K)^2 = constant = Lorenz's\,number$$

where k = thermal conductivity, k_e = electrical conductivity,

T = absolute temperature, while the thermal conductivity for all metals (pure, alloyed, solid or liquid) is given as

$$k = 2.61 \; x \; 10^{-8} \frac{T}{\rho_e} - \frac{2 \; x \; 10^{-17} T^2}{Cp.\rho.\rho_e^2} + 97 \frac{Cp.\rho^2}{M.T}, \qquad \frac{W}{m.K}$$

where T = absolute temperature, $^{\circ}$K, ρ_e = electrical resistivity, ohm – cm, ρ = density, g/cc, Cp = specific heat, cal/g.C, M = average atomic weight, g/g. atom

Determine the thermal conductivity of pure aluminium and a brass consisting of 70% copper and 30% nickel, at 293K, using these equations.

Answer: From standard definitions:

Electrical Resistance in $Ohms = R = \frac{\rho_e L}{A}$

Electrical Resistivity in $Ohm\text{-}cm = \rho_e = R\frac{A}{L}$

Electrical conductivity in $Ohm^{-1}cm^{-1} = k_e = \frac{1}{\rho_e}$

The data required for the calculations are obtained from Welty (1978) and from the CRC Handbook (1970) as follows

Metal	Density, ρ, g/cc	Sp. Heat, Cp, kJ/kg.K	AtomicWeight, M	Electrical Resistivity, ρ_e, Ohm-cm
Aluminium	2.700	0.938	27	2.655 x 10^{-6}
Pure Copper	8.890	0.385	64	1.673 x 10^{-6}
Pure Nickel	8.910	0.465	59	6.850 x 10^{-6}
Brass	8.520	0.381	62.5	35 x 10^{-6}

Since $Power = V \ x \ I = V \ x \frac{V}{R} = \frac{V^2}{R}, Watts \ (W)$

then $R, Ohms = \frac{(Volts)^2}{Power}$ with units of $\frac{V^2}{W}$.

Note that W has units of $\frac{(Volt)^2}{Ohm}$.

Recall, also, the definition of electrical conductivity as $k_e = \frac{1}{\rho_e}, Ohm^{-1}cm^{-1}$

Thus, the units of $\frac{k}{k_e T}$ in the Wiederman, Franz, Lorenz equation will be

$$\frac{W/cm.K}{K/Ohm.cm} = \frac{W}{cm.K} . \frac{Ohm.cm}{K} = \frac{V^2/Ohm}{cm.K} . \frac{Ohm.cm}{K} = \frac{V^2}{K^2}$$

For Aluminium

$$k_e = \frac{1}{\rho_e} = \frac{1}{2.655 \ x \ 10^{-6}} = 3.7665 \ x \ 10^5, W/V^2cm$$

Using the Weidman, Franz and Lorenz equation,

$$k = 29 \ x \ 10^{-9} x \ k_e T , \frac{V^2}{K^2} \ x \ \frac{W}{V^2 cm} x \ K \ x \ 100 \ \frac{cm}{m}$$
$$= 29 \ x \ 10^{-7} x \ k_e T , \frac{W}{m.K}$$

At T = 293 K, the Weidman, Franz and Lorenz equation gives

$$k_{Al} = 29 \ x \ 10^{-7} x 3.7665 \ x \ 10^5 \ x \ 293 = 320.0 \ \frac{W}{m.K}$$

The actual value is 228 W/m.K; that is an error of +40%.

For Brass

$$k_e = \frac{1}{\rho_e} = \frac{1}{35 \ x \ 10^{-6}} = 2.8571 \ x \ 10^4, W/V^2cm$$

At T = 293 K, the Weidman, Franz and Lorenz equation gives

$$k_{Br} = 29 \times 10^{-7} \times 2.8571 \times 10^4 \times 293 = 24.3 \frac{W}{m.K}$$

The actual value is 107 W/m.K; that is an error of -77%.

These two examples, one for a pure metal and another for an alloy, show that the Weidman, Franz and Lorenz equation is unsuitable for predicting the thermal conductivity of metals and alloys.

Using the equation applicable to all metals

$$k = 2.61 \times 10^{-8} \frac{T}{\rho_e} - \frac{2 \times 10^{-17}}{Cp.\rho.\rho_e^2} T^2 + 97 \frac{Cp.\rho^2}{M.T}, \frac{W}{cm.K}$$

<u>For Aluminium</u>, at T = 293 K

$$Cp = 0.938 \frac{kJ}{kg.K} = 0.938 \frac{J}{g.K} = \frac{0.938 \ J/gK}{4.187 \ J/Cal} = 0.224 \frac{Cal}{g.K}$$

Then

$$k_{Al} = 2.61 \times 10^{-8} \frac{293}{2.655 \times 10^{-6}} - \frac{2 \times 10^{-17}}{0.224 \times 2.7 \times (2.655 \times 10^{-6})^2} \times (293)^2$$
$$+97 \frac{0.224 \times (2.7)^2}{27 \times 293}$$

$$= 2.8803 - 0.4027 + 0.0200 = 2.4976 \frac{W}{cm.K}$$

Actual value = 2.28 W/cm.K ie. 9.5% in error.

<u>For Brass</u>, at T = 293K

$$Cp = 0.381 \frac{kJ}{kg.K} = 0.381 \frac{J}{g.K} = \frac{0.381 \ J/gK}{4.187 \ J/Cal} = 0.091 \frac{Cal}{g.K}$$

Then

$$k_{Br} = 2.61 \; x \; 10^{-8} \; \frac{293}{35 \; x \; 10^{-6}} - \frac{2 \; x \; 10^{-17}}{0.091 \; x \; 8.52 \; x \; (35 \; x \; 10^{-6})^2} x \; (293)^2$$

$$+97 \; \frac{0.091 \; x \; (8.52)^2}{62.5 \; x \; 293}$$

$$= 0.2185 - 0.0018 + 0.0350 = 0.2517 \; \frac{W}{cm.\,K}$$

Actual value = 1.07 W/cm.K, ie +76.5 %, very much, in error.

Example 3.17: The thermal conductivity of wood, for heat flow normal to the grain, is given by Green (1984) as

$$k = \rho(0.1159 + 0.00233M) + 0.01375, BTU/h.\,ft.\,F$$

where ρ, the density of the wood, g/cc, is the weight when dry divided by the volume when green and M is the moisture content less than 40% by weight.

Determine the thermal conductivity of oak for heat flow normal to the grain given that $\rho = 0.820$, g/cc and M = 15%

Answer: Using the values given

k = 0.820 (0.1159 + 0.00233 x 15) + 0.01375 = 0.1375 Btu/h.ft F
  = 0.1375 x 1.7307 = 0.238 W/m.K

Actual value = 0.21 W/m.K; that is 13% in error.

Example 3.18: The Russell equation for the thermal conductivity of porous media (solid and liquid or solid and gas) is given by:

$$\frac{k_m}{k_{cont}} = \frac{v\rho^{2/3} + 1 - \rho^{2/3}}{v(\rho^{2/3} - \rho) + 1 - \rho^{2/3} + \rho}$$

where

$$\rho = \frac{\rho_{solid} - \rho_m}{\rho_{solid} - \rho_{gas/liquid}}$$

Subscript **m** represents the composite material, **cont** represents the continuous phase and $v = k_{solid}/k_{cont}$. Determine the thermal conductivity of a sand bed in air given that $\rho_{sand} = 2{,}519$ kg/m³, $\rho_{air} = 1.22$ kg/m³ and that the voidage of the bed is 26%. The bulk density of the bed is 802 kg/m³.

Answer: From Welty (1978) and Perry *et al* (1963),

$$k_{sand} = 328.8 \text{ W/m.K}; \quad k_{air} = 2.546 \text{ x } 10\text{-}2 \text{ W/m.K}$$

$$\rho = \frac{\rho_{solid} - \rho_m}{\rho_{solid} - \rho_{gas/liquid}} = \frac{2{,}519 - 802}{2{,}519 - 1.22} = 0.68$$

$$v = \frac{k_{solid}}{k_{air}} = \frac{328.8}{2.546 \text{ x } 10^{-2}} = 12{,}914$$

$$\frac{k_m}{k_{cont}} = \frac{12{,}914 \text{ x } (0.68)^{2/3} + 1 - (0.68)^{2/3}}{12{,}914((0.68)^{2/3} - 0.68) + 1 - (0.68)^{2/3} + 0.68}$$

$$= \frac{9{,}986.4}{1{,}205.78} = 8.282$$

Since air is the continuous phase, $k_{cont} = k_{air}$

$$k_m = 8.282 \text{ x } k_{air} = 8.282 \text{ x } 2.546 \text{ x } 10^{-2}$$
$$= 0.211 \text{ W/m.K } Ans.$$

References For Chapter Three

1   Bird, R.B; Stewart, W.E; Lightfoot, E.N; *Transport Phenomena*, Chapters 8 – 14; Wiley Int'l Edition, NY., USA 1960.

2   Carslaw, H.S and Jaeger, J.C; *Conduction of Heat in Solids*, Clarendon Press, Oxford, UK, 1959.

3   Charm S.E; *The Fundamentals of Food Engineering*, Chapter 4, 2nd Edn, AVI Publishing Co; Westport, Connecticut, USA, 1971.

4   Heldman, D.R; *Food Process Engineering*, Chapter 3, AVI Publishing, Westport, Connecticut, USA, 1975.

5   http://en.wikipedia.org/wiki/Heat transfer

6   http://web.mit.edu/16.unified/www/FALL/thermodynamics/not es/node118.html

7   http://www.classroom-energy.org/teachers/energy_tour/pg1-5. html

8   Perry R.H, Chilton C.H and Kirkpatrick S.D; *Perry's Chemical Engineers' Handbook*, 4th Edition; McGraw - Hill Book Co; New York, USA, 1963.

9   Perry R.H and Green D; *Chemical Engineers' Handbook*. 6th Edition; McGraw – Hill Book Co; New York, USA, 1984.

10 Smith B.D: *Design of Equilibrium Stage Processes*, McGraw - Hill Book Company, New York, USA, 1963.

11 Classroom-energy.org, 2009, Wikipedia, 2009

12 International Critical Tables Vol. 5, p 227

13 CRC Handbook (1970)

CHAPTER FOUR
STEADY STATE, ONE-DIMENSIONAL, HEAT CONDUCTION

Example 4.01: Why is it that, even though real life is in three dimensions, we are interested in one dimensional heat conduction?

Answer: Analysing one dimensional heat conduction is of interest and more popular than analysing two or three dimensional heat conduction because it is easier to do and, more importantly, can be applied to a large number of practical situations. In addition, by the mathematical methods of superposition of solutions, the so called product solutions may be obtained in which one dimensional solutions are combined to give results for two or three dimensions.

Example 4.02: What is steady state heat conduction?

Answer: By steady state heat conduction, we mean that, during the heat conduction process, the temperature does not change with time. That is

$$\frac{\partial T}{\partial t} = 0 \tag{4.01}$$

Example 4.03: How important is steady state heat conduction?

Answer: Steady state heat conduction is important in many commercial applications and situations because most product manufacturers prefer to operate at steady state because it ensures consistent product quality.

Example 4.04: Derive the expression for the one dimensional and steady state heat conduction through an isotropic flat plate or wall.

Answer: The derivation of the expression for the one dimensional, steady state, heat conduction is, probably, the simplest and most straightforward derivation which relates the, mathematically, postulated Fourier's first law of heat conduction to actual real life heat conduction.

Consider, for example, the schematic, shown below, which illustrates heat conduction through an isotropic flat plate or wall. The temperatures at the surfaces of the wall, the wall thickness and the heat flux perpendicular and through the wall are represented by the standard symbols in the subject.

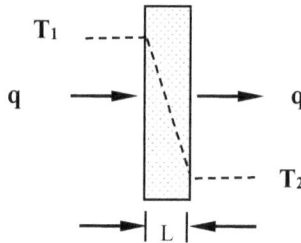

Because the heat transfer is steady state with no internal generation or consumption of heat energy, the Laplace's equation is the equation most suited to the solution of the problem.

In one dimension, in the x-direction, the Laplace equation is

$$\frac{\partial^2 T}{\partial x^2} = 0 = \frac{d^2 T}{dx^2} \qquad (4.02)$$

Note that we have gone from partial to total differential equations because other dimensions have no effect on the conduction. Integrating once

$$\frac{dT}{dx} = C, \quad \text{a constant} \tag{4.03}$$

Integrating again

$$\int_{T_1}^{T_2} dT = \int_0^L C dx \tag{4.04}$$

That is

$$T_2 - T_1 = C.L \tag{4.05}$$

If a general T, at any general x, replaces the particular T_2 at x_2, equation (4.05) can be expressed more generally, as

$$T = T_1 + C(x - x_1) \tag{4.06}$$

From equations (4.03) and (4.05), we find that

$$C = \frac{dT}{dx} = \frac{T_2 - T_1}{L} \tag{4.07}$$

Substituting equation (4.07) into the equation for Fourier's first law of heat conduction, we get that

$$q = -k\frac{dT}{dx} = \frac{k}{L}(T_1 - T_2) \tag{4.08}$$

where q is the heat flux through the plate or wall. This is the expression for the one dimensional and steady state heat conduction through an isotropic flat plate or wall

Example 4.05: Calculate the rate of heat transfer across one square meter of a brick wall 50 mm thick whose thermal conductivity, $k = 0.93$ W/m.K, if the two faces are at 30 C and 5 C respectively.

Answer: From equation (4.08) above

$$q = \frac{k}{L}(T_1 - T_2) = \frac{0.93}{0.05}(303 - 278), \frac{W}{m.K}.\frac{1}{m}.K = \frac{0.93}{0.05}x25, \frac{W}{m^2}$$

$$= 465 \ W/m^2 \ Ans.$$

Example 4.06: A boiler is to be insulated such that the heat loss will not exceed 2.4 kW per square meter of wall area. What thickness of asbestos is required if the inner and outer surfaces of the insulation are to be 1090K and 480K respectively?

Answer: The physical situation can be shown using a section of the insulated boiler wall.

Boiler Wall   Asbestos Insulation

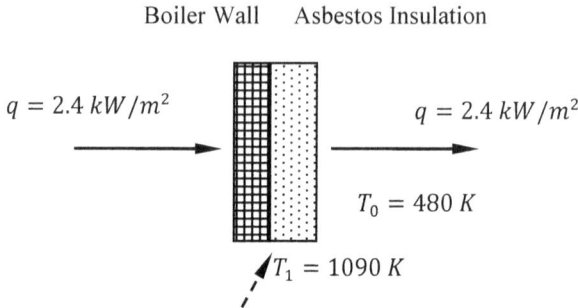

$q = 2.4 \ kW/m^2$

$q = 2.4 \ kW/m^2$

$T_0 = 480 \ K$

$T_1 = 1090 \ K$

By Fourier's first law of heat conduction

$$q = -k\frac{dT}{dx} \qquad (1)$$

which for the case above, reduces to

$$q = -k\frac{T_0 - T_L}{L} \qquad (2)$$

From data tables, for asbestos, k = 0.214 W/m.K at 573 K and 0.188 W/m.K at 373K.

Arithmetic mean temperature of asbestos insulation

$$= \frac{1090 + 480}{2} = 785 \ K$$

If we assume linear variation of k with temperature

$$\frac{k_{785} - k_{573}}{785 - 573} = \frac{k_{573} - k_{373}}{573 - 373}$$

Then

$$k_{785} = k_{573} + (k_{573} - k_{373})\frac{785 - 573}{573 - 373}$$

$$k_{785} = 0.214 + (0.214 - 0.186)\frac{785 - 573}{573 - 373} = 0.242 \ W/mK \quad (3)$$

Substituting (3) and given values of q, T_L, and T_0 in equation (2) we get

$$2400, \frac{W}{m^2} = -0.242.\frac{(480-1090)}{L}, \frac{W}{m.K}.\frac{K}{m}$$

From which

$$L = \frac{0.242 \ x \ 610}{2400} = 0.062 \ m. \qquad Ans$$

Example 4.07: Derive the expression for the one dimensional, steady state heat conduction through an isotropic flat plate or wall when the thermal conductivity, k, varies with temperature

Answer: If the thermal conductivity of an isotropic solid varies with temperature in a linear manner such that $k = k_0 [1 + \beta(T - T_0)]$, the heat flux may be determined as follows.

Since

$$q = -k\frac{dT}{dx}, \qquad then \qquad -kdT = qdx$$

Thus

$$\int_{T_1}^{T_2} -kdT = \int_{T_1}^{T_2} -k_0[1 + \beta(T - T_0)] \ dT = \int_{0}^{L} qdx$$

That is

$$\left[k_0(T_2 - T_1) + k_0\frac{\beta}{2}(T_2^2 - T_1^2) - k_0\beta T_0(T_2 - T_1) \right] = qL$$

This simplifies to

$$qL = -k_o(T_2 - T_1)\left[1 + \frac{\beta}{2}(T_2 + T_1) - \beta T_o\right] \qquad (4.09)$$

If we define an average $k = k_m$, by comparison with equation (4.08) such that

$$k_m = k_o\left[1 + \frac{\beta}{2}(T_2 + T_1) - \beta T_o\right] \qquad (4.10)$$

Then

$$qL = -k_m(T_2 - T_1) \quad or \quad q = \frac{k_m}{L}(T_1 - T_2) \qquad (4.11)$$

This equation is similar to that for the heat flux through a flat plate or wall and in which the thermal conductivity is independent of temperature, except that a mean thermal conductivity, dependent on temperature, now replaces the temperature-independent thermal conductivity, k.

Example 4.08: Calculate the heat flux in the previous example if

$$k = 0.93[1 + 0.0054T], W/mK$$

Answer: The form of the above equation for thermal conductivity implies that $T_0 = 0$. Thus, from equation (4.10)

$$k_m = k_o\left[1 + \frac{\beta}{2}(T_2 + T_1) - \beta T_o\right]$$
$$= 0.93\left[1 + \frac{0.0054}{2}(303 + 278)\right]$$
$$= 2.389 \ W/mK$$

Hence from equation (4.11)

$$q = \frac{2.389}{0.05} \ x25 = 1194.5 \ W/m^2. \quad Ans$$

Example 4.09: Derive the expression for heat conduction through a composite wall where each component wall is isotropic with constant thermal conductivity.

Answer: Such a wall may be that of an ordinary house in which the component walls are the inner coat of paint or plaster, the structural wall of brick or concrete and an outer coat of paint or plaster.

Such a composite wall can, also, be that of a furnace consisting of an inner fireclay brick wall, a structural metal or building brick wall and an outer insulating or building brick wall. We assume, here, that brick, concrete and paint are isotropic materials (in practice, they may not be).

Whatever the composition of the wall, it is assumed that there is perfect contact between the surfaces of the component walls. That is

$$q_1 = q_2 = q_3 = q = constant \qquad (4.12)$$

The interest is, usually, to determine the heat flux through the composite wall if the surface temperatures are known or the surface temperatures if the heat flux is known. Also, the relative resistance to heat transfer, of each of the components, is, sometimes, of interest.

Thus, a small section of such a composite wall, made up of three components, may be illustrated as shown in the diagram.

For conduction through a single wall, we know from equation (4.08) that for wall number 1,

$$q = \frac{k_1}{L_1}(T_1 - T_2) \quad \text{or} \quad T_1 - T_2 = \frac{qL_1}{k_1} \qquad (4.11a)$$

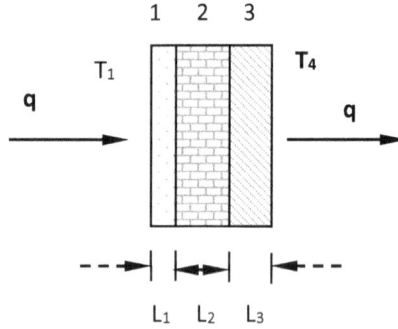

Similarly, for walls numbers 2 and 3,

$$T_2 - T_3 = \frac{qL_2}{k_2} \qquad (4.11b)$$

$$T_3 - T_4 = \frac{qL_3}{k_3} \qquad (4.11c)$$

Adding equations (4.11a), (4.11b) and (4.11c), we find that

$$T_1 - T_4 = q\left[\frac{L_1}{k_1} + \frac{L_2}{k_2} + \frac{L_3}{k_3}\right]$$

From which we get that

$$q = \frac{T_1 - T_4}{\frac{L_1}{k_1} + \frac{L_2}{k_2} + \frac{L_3}{k_3}} \qquad (4.13)$$

This may be generalised, for n walls, as

$$q = \frac{T_1 - T_{n+1}}{\sum_{i=1}^{n}\left(\frac{L_i}{k_i}\right)} \qquad (4.14)$$

The total flow of heat, Q = q.A, is given by

$$q.A = \frac{T_1 - T_{n+1}}{\sum_{i=1}^{n}\left(\frac{L_i}{A_i.k_i}\right)} \qquad (4.15)$$

$\frac{L_i}{A_i.k_i}$ is *thermal resistance* $\qquad (4.16)$

A is the area normal to the direction of the heat flux. For plane parallel walls, all the A_i are equal.

Example 4.10: A furnace wall is constructed with 0.23 m of firebrick, 0.13 m of insulating brick and 0.23 m of building brick. The inside temperature is 930 C and the outside is at 55°C. If the thermal conductivities are, for firebrick, 1.385 W/m.K, insulating brick, 0.208 W/m.K and for ordinary brick, 0.693 W/m.K, find the heat loss per unit area and the temperature at the junction of firebrick and insulating brick.

Answer

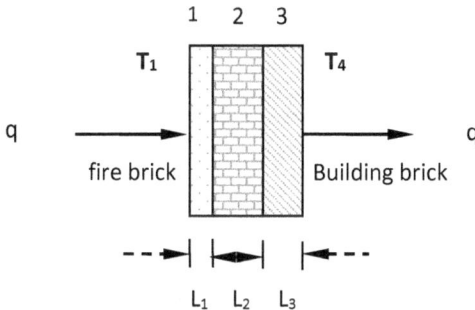

Assume that the firebrick surface temperature is the same as the furnace temperature = $T_1 = 930°C = 1203$ K. $T_4 = 55$ C $= 328$ K

From equation (4.14),

$$q = \frac{T_1 - T_{n+1}}{\sum_{i=1}^{n}\left(\frac{L_i}{k_i}\right)} = \frac{1203 - 328}{\frac{0.23}{1.385} + \frac{0.13}{0.208} + \frac{0.23}{0.693}}$$

$$= \frac{875}{0.166 + 0.625 + 0.332} \cdot \frac{K}{1} \cdot \frac{W}{m.\,m.\,K} = 779.16 \; W/m^2 \quad Ans$$

To find the temperature of the interface between firebrick and insulating brick walls, we shall use equation (4.11a). Thus

$$T_1 - T_2 = \frac{qL_1}{k_1} = 1203 - T_2 = \frac{779.16 \; x \; 0.23}{1.385} = 129.39$$

Therefore $T_2 = 1203 - 129.39 = 1073.61 \; K \; (800.61 \; C) \quad Ans$

Example 4.11: A composite wall is constructed of 0.01m of aluminium, 0.01m of corkboard, and 3mm of plastic. The outside temperature of the aluminium is 395K and the plastic surface is held at 300K. Determine the temperatures on either side of the corkboard and the heat flux across the composite wall under these conditions.

You are given that for aluminium, k_{Al} = 230 W/m.K; corkboard, k_C = 0.043 W/m.K; plastic, k_p = 2.2 W/m.K

Answer: The physical situation may be represented as shown below.

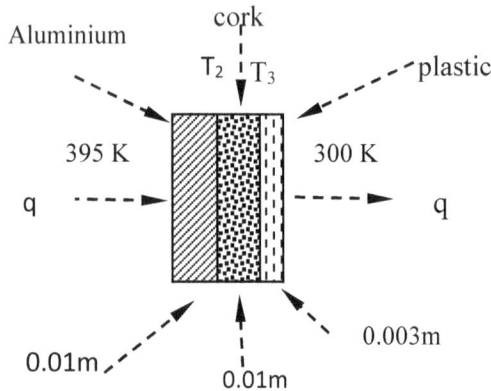

For aluminium:

$$q = -k\frac{dT}{dx} = -k_{Al}\frac{395 - T_2}{0.01}$$

or

$$395 - T_2 = \frac{q \times 0.01}{k_{Al}} \qquad (1)$$

Similarly, for corkboard

$$T_2 - T_3 = \frac{q \times 0.01}{k_C} \qquad (2)$$

For Plastic

$$T_3 - 300 = \frac{q \times 0.003}{k_P} \qquad (3)$$

Adding equations (1), (2) and (3)

$$395 - 300 = 95 = q\left[\frac{0.01}{k_{Al}} + \frac{0.01}{k_C} + \frac{0.003}{k_P}\right] = q\left[\frac{0.01}{230} + \frac{0.01}{0.043} + \frac{0.003}{2.2}\right]$$

$$= q(4.348 \times 10^{-5} + 0.2326 + 1.3636 \times 10^{-3}) = q \times 0.234$$

This gives $q = 406\,W/m^2$ Ans.

Adding equations (2) and (3)

$$T_2 - 300 = q\left[\frac{0.01}{k_C} + \frac{0.003}{k_P}\right] = q(0.2326 + 1.3636 \times 10^{-3})$$

$$= 406 \times 0.23396 = 94.99$$

Hence $T_2 = 394.99\,K.\,Ans$

Adding equations (1) and (2)

$$395 - T_3 = q\left[\frac{0.01}{k_{Al}} + \frac{0.01}{k_C}\right] == 406(4.348 \times 10^{-5} + 0.2326)$$

$$= 406 \times 0.2326438 = 94.45$$

Hence $T_2 = 300.55\ K.\ Ans$

Example 4.12: A furnace wall consists of 0.106 m layer of fireclay brick and 0.635 cm thickness of mild steel on its outer surface. The temperature inside the furnace is 1500 K while outside air is at 303 K. If the heat transfer coefficients on the inside and outside wall surfaces are 5110 W/m²K and 45 W/m²K respectively, determine the heat lost per unit area from the furnace and the temperature at each surface and at the brick steel interface.

Take the thermal conductivity for steel k_S = 39.6 W/mK and for firebrick k_f = 1.11 W/mK.

Answer: Given

T_i =1500 K	h_i = 5110 W/m²K	$k_{fireclay}$ = 1.11 W/mK
T_0 = 303 K	h_0 = 45 W/m²K	k_S = 39.6 W/mK

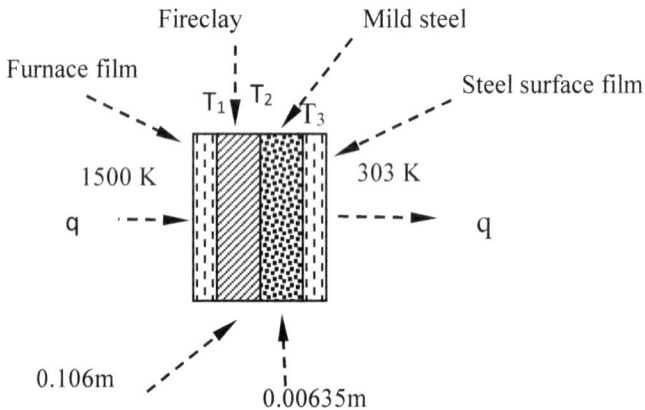

Heat loss from furnace to fireclay brick through the furnace film

$$q = h_i(1500 - T_1)$$

or

$$1500 - T_1 = \frac{q}{h_1} \tag{1}$$

Similarly, from fireclay brick to mild steel

$$T_1 - T_2 = \frac{q \cdot x_{fireclay}}{k_{fireclay}} \tag{2}$$

From mild steel to air

$$T_2 - T_3 = \frac{q \cdot x_s}{k_s} \tag{3}$$

Across the air or mild steel surface film

$$T_3 - 303 = \frac{q}{h_0} \tag{4}$$

Adding equations (1), (2), (3) and (4)

$$1500 - 303 = q \left[\frac{1}{h_i} + \frac{x_{fireclay}}{k_{fireclay}} + \frac{x_s}{k_s} + \frac{1}{h_0} \right]$$

$$= q \left[\frac{1}{5110} + \frac{0.106}{1.11} + \frac{0.00635}{39.6} + \frac{1}{45} \right]$$

$$1197 = q(1.356 \; x \; 10^{-4} + 0.0955 + 1.604 \; x \; 10^{-4} + 0.0222)$$

$$= q \; x \; 0.117996$$

This gives

$$q = \frac{1197}{0.117996} = 10,144.4 \; W/m^2 \quad Ans$$

From equation (1):

$$T_1 = 1500 - \frac{q}{h_1} = 1500 - \frac{10144.4}{5110} = 1498 \; K$$

From equations (1) and (2)

$$1500 - T_2 = q \left[\frac{1}{h_i} + \frac{x_{fireclay}}{k_{fireclay}} \right]$$

$$= 10{,}144.4 \, x \, (1.356 \, x \, 10^{-4} + 0.0955)$$

Hence

$$T_2 = 1500 - 10{,}144.4 \, x \, 0.0956356 = 529.8 \, K$$

From equations (1), (2) and (3)

$$1500 - T_3 = q \left[\frac{1}{h_i} + \frac{x_{fireclay}}{k_{fireclay}} + \frac{x_s}{k_s} \right]$$

$$= 10{,}144.4 \, x \, (1.356 \, x \, 10^{-4} + 0.0955 + 1.604 \, x \, 10^{-4}) = 971.79$$

Thus $T_3 = 528.2 \, K$

Also, from equation (4)

$$T_3 - 303 = \frac{10{,}144.4}{45}$$

That is $T_3 = 528.4 \, K$

Example 4.13: Derive the expression for steady state, one dimensional heat conduction along the radius of a long hollow cylinder.

Answer: Because the object is a cylinder, it makes sense that the cylindrical co-ordinate system would be the most appropriate system to use. Conduction in the radial direction is industrially and practically important because industrial heating with steam or cooling with water takes place in pipes. It is, therefore, the one dimension chosen here for analysis.

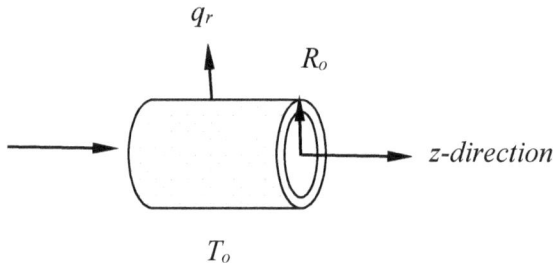

The steady state Laplace's equation, in cylindrical co-ordinates, in the r-direction, is

$$\frac{\partial^2 T}{\partial r^2} + \frac{1}{r}\frac{\partial T}{\partial r} = 0 = \frac{1}{r}\frac{\partial}{\partial r}\left(r\frac{\partial T}{\partial r}\right) \qquad (4.17)$$

This is equivalent to

$$\frac{1}{r}\frac{d}{dr}\left(r\frac{dT}{dr}\right) = 0 \qquad (4.17a)$$

On integration, it gives

$$r\frac{dT}{dr} = C, \qquad a\ constant \qquad (4.17b)$$

Integrating equation (4.17b)

$$\int_{T_i}^{T_o} dT = C \int_{R_i}^{R_o} \frac{dr}{r}$$

we get

$$T_o - T_i = C.ln\left(\frac{R_o}{R_i}\right)$$

Thus

$$C = \frac{T_o - T_i}{ln\left(\frac{R_o}{R_i}\right)} \qquad (4.18)$$

and for any **r** and T

$$T - T_i = \frac{T_o - T_i}{ln\left(\frac{R_o}{R_i}\right)}.ln\left(\frac{r}{R_i}\right) \qquad (4.19)$$

This is the equation for the temperature distribution.

From equations (4.17b) and (4.18)

$$\frac{dT}{dr} = \frac{C}{r} = \frac{1}{r}.\frac{T_o - T_i}{ln\left(\frac{R_o}{R_i}\right)} \qquad (4.20)$$

The heat flux in the r-direction is given, therefore, as

$$q_r = -k\frac{\partial T}{\partial r} = \frac{k}{r}\cdot\frac{T_i - T_o}{\ln\left(\frac{R_o}{R_i}\right)} \tag{4.21}$$

Note, from equation (4.21), that the heat flux along the radius of the cylinder varies with the radius. Thus at the outermost radius of the hollow cylinder

$$q_{R_o} = \frac{k}{R_o}\cdot\frac{T_i - T_o}{\ln\left(\frac{R_o}{R_i}\right)} \tag{4.21a}$$

while at the innermost radius of the hollow cylinder

$$q_{R_i} = \frac{k}{R_i}\cdot\frac{T_i - T_o}{\ln\left(\frac{R_o}{R_i}\right)} \tag{4.21b}$$

For a given length of cylinder, L, of circumferential area A, total heat flow, Q, is given by

$$Q = q_r.A = q_r.2\pi rL = \frac{2\pi kL(T_i - T_o)}{\ln\left(\frac{R_o}{R_i}\right)} = constant \tag{4.22}$$

Example 4.14: Derive the expression for steady state, one dimensional heat conduction along the radius of a long hollow cylinder when k varies with temperature.

Answer: If, for the hollow cylinder, $k = k_o [1+\beta(T-T_o)]$, an analysis, similar to that for a flat plate can be done. The net result is

$$q_r = \frac{k_m}{r}\cdot\frac{T_i - T_o}{\ln\left(\frac{R_o}{R_i}\right)} \tag{4.23}$$

and

$$Q = \frac{2\pi k_m L(T_i - T_o)}{\ln\left(\frac{R_o}{R_i}\right)} = constant \qquad (4.24)$$

where

$$k_m = k_o\left[1 + \frac{\beta}{2}(T_i + T_o) - \beta T_o\right]$$

Example 4.15: Derive the expression for steady state, one dimensional heat conduction in the radial direction in a long, composite, hollow cylinder

Answer: This may be a pipe, painted or insulated on the outside or inside, so that multiple diameters or radii can be identified as shown in the diagram. As in the composite plane wall, it is assumed that perfect physical contact exists between the surfaces. Since

$$Q = \frac{2\pi k L(T_i - T_o)}{\ln\left(\frac{R_o}{R_i}\right)} = constant \qquad from \ (4.24)$$

for the figure shown

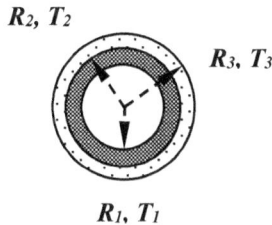

R_2, T_2

R_3, T_3

R_1, T_1

$$T_2 - T_1 = \frac{Q\ln\left(\frac{R_2}{R_1}\right)}{2\pi k_1 L} \qquad (4.24a)$$

$$T_3 - T_2 = \frac{Qln\left(\frac{R_3}{R_2}\right)}{2\pi k_2 L} \qquad (4.24b)$$

Adding equations (4.24a) and (4.24b) together

$$T_3 - T_1 = \frac{Q}{2\pi L}\left[\frac{1}{k_1}ln\left(\frac{R_2}{R_1}\right) + \frac{1}{k_2}ln\left(\frac{R_3}{R_2}\right)\right] \qquad (4.25)$$

Example 4.16: A pipe carrying saturated steam at 488K passes through a room in which the air temperature is 300K. Determine the heat loss per meter of pipe if the resistances on the inside and outside of the pipe are 0.1036 m²K/kW and 25 m²K/kW respectively. Take the ID of pipe to be 0.059 m, OD to be 0.073 m and k = 42.9 W/mK.

Answer: The cross-section of the pipe may be shown below.

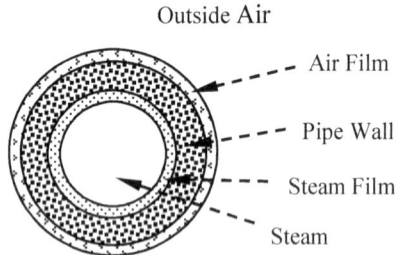

For heat transfer in a cylinder $Q = q.A = -2\pi rLk\frac{dT}{dr}$

Rearranging and integrating:

$$\int_{R_i}^{R_o}\frac{dr}{r} = -\frac{2\pi kL}{Q}\int_{T_i}^{T_o}dT$$

$$ln\left(\frac{R_o}{R_i}\right) = -\frac{2\pi Lk}{Q}(T_o - T_i) = \frac{2\pi kL}{Q}(T_i - T_o)$$

Hence

$$T_i - T_o = \frac{Q}{2\pi kL} \ln\left(\frac{R_o}{R_i}\right) \tag{1}$$

For the inner fluid steam

$$Q = A.h.\Delta T = 2\pi R_i L h_i (T_s - T_i)$$

or

$$T_s - T_i = \frac{Q}{2\pi R_i L h_i} \tag{2}$$

For the pipe

$$T_i - T_o = \frac{Q}{2\pi kL} \ln\left(\frac{R_o}{R_i}\right) \tag{3}$$

For the outer film

$$T_o - T_{air} = \frac{Q}{2\pi R_o L h_o} \tag{4}$$

Adding equations (2), (3) and (4)

$$T_s - T_{air} = \frac{Q}{2\pi L}\left[\frac{1}{R_i h_i} + \frac{1}{k}\ln\left(\frac{R_o}{R_i}\right) + \frac{1}{R_o h_o}\right]$$

Rearranging in order to use the given values directly

$$T_s - T_{air} = \frac{Q}{2\pi L}\left[\frac{2}{D_i}\cdot\frac{1}{h_i} + \frac{1}{k}\ln\left(\frac{D_o}{D_i}\right) + \frac{2}{D_o}\cdot\frac{1}{h_o}\right]$$

Substituting the given values

$$488 - 300 = \frac{Q}{2\pi L}\left[\frac{2}{0.059}x0.1036 + \frac{1}{0.0429}\ln\left(\frac{0.073}{0.059}\right) \\ + \frac{2}{0.073}x25\right]$$

That is

$$188 = \frac{Q}{\pi L}[1.7559 + 2.4816 + 342.4658]$$

Hence

$$\frac{Q}{L} = \frac{188 \, x \, \pi}{346.7033} = 1.7035 \; kW/m \qquad Ans.$$

Example 4.17: A 0.0737m I.D, 0.0889m O.D mild steel steam line is to carry 395 K steam through a tunnel, where the temperature may get as high as 310 K. In order that the surface temperature may never exceed 335K, how much 85% magnesia insulation must be used to cover the steam pipe? The values of h applying at the inside and outside surfaces are 4.8 kW/m^2.K and 68 W/m^2.K.

Answer: The physical situation is equivalent to that of heat conduction along the radii of four concentric composite cylinders illustrated as follows

$T_1 = 495$ K

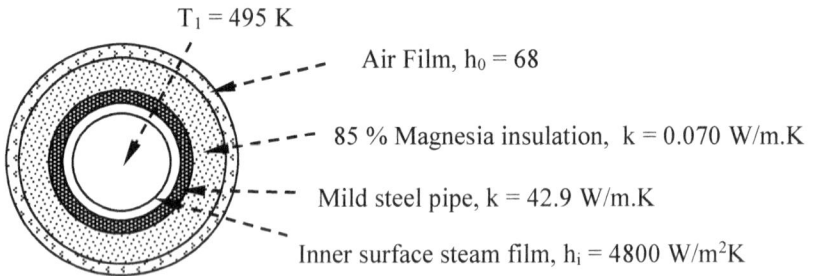

Air Film, $h_o = 68$

85 % Magnesia insulation, k = 0.070 W/m.K

Mild steel pipe, k = 42.9 W/m.K

Inner surface steam film, $h_i = 4800$ W/m^{2}K

$T_{tunnel} = 310$ K, $T_{surface} = T_0 = 335$ K

For heat transfer across an area, which varies with distance along the direction of heat transfer, $Q = qA = $ constant.

By Fourier's first law of heat conduction

$$q = -k\frac{dT}{dr}$$

and A=2πrL, where L is length along the cylinder, from which we get that

$$Q = q.A = -2\pi r Lk \frac{dT}{dr} \qquad (1)$$

For any hollow cylinder of inner and outer radius R_i and R_0 we can integrate equation (1) as

$$\int_{R_i}^{R_o} \frac{dr}{r} = -\frac{2\pi kL}{Q} \int_{T_i}^{T_o} dT$$

to get

$$\ln\left(\frac{R_o}{R_i}\right) = -\frac{2\pi Lk}{Q}(T_o - T_i) = \frac{2\pi kL}{Q}(T_i - T_o)$$

Hence, per unit length

$$T_i - T_o = \frac{Q}{2\pi k} \ln\left(\frac{R_o}{R_i}\right) \qquad (2)$$

For the steam film inside the pipe, per unit length,

$$Q = h_i.A_i.\Delta T_i = 2\pi R_i h_i(T_i - T_w)$$

or

$$T_i - T_w = 395 - T_w = \frac{Q}{2\pi R_i h_i} \qquad (3)$$

For the steel pipe

$$T_w - T_s = \frac{Q}{2\pi k_s} \ln\left(\frac{R_o}{R_i}\right) \qquad (4)$$

For the insulation

$$T_s - T_o = T_s - 335 = \frac{Q}{2\pi k_{ins}} \ln\left(\frac{R_{o,ins}}{R_{i,ins}}\right) \qquad (5)$$

For the air film

$$T_o - T_{air} = 335 - 310 = \frac{Q}{2\pi R_o h_o} \qquad (6)$$

Adding equations (3), (4), (5) and (6)

$$395 - 310 = \frac{Q}{2\pi}\left[\frac{1}{R_i h_i} + \frac{1}{k_s}\ln\left(\frac{R_o}{R_i}\right) + \frac{1}{k_{ins}}\ln\left(\frac{R_{o,ins}}{R_{i,ins}}\right) + \frac{1}{R_{o,ins}h_o}\right]$$

Substituting the given values

$$85 = \frac{Q}{2\pi} \left[\frac{1}{0.03685 \times 4800} + \frac{1}{42.9} \ln \left(\frac{0.0889}{0.0737} \right) + \frac{1}{0.070} \ln \left(\frac{R_{o,ins}}{R_{i,ins}} \right) \right.$$
$$\left. + \frac{1}{0.04445 \times 68} \right]$$

That is

$$\frac{534.0708}{Q} = 0.005654 + 0.004371 + 14.2857 \ln \left(\frac{R_{o,ins}}{R_{i,ins}} \right)$$
$$+ 0.3308 = 0.3408 + 14.2857 \ln \left(\frac{R_{o,ins}}{R_{i,ins}} \right)$$

or

$$\ln \left(\frac{R_{o,ins}}{R_{i,ins}} \right) = \frac{37.3850}{Q} - 0.02386 \qquad (7)$$

From equation (6)

$$335 - 310 = \frac{Q}{2\pi R_o h_o} = \frac{Q}{2\pi \times 0.04445 \times 68}$$

This gives Q = 474.789 W.

Substituting this value of Q in equation (7)

$$\ln \left(\frac{R_{o,ins}}{R_{i,ins}} \right) = \frac{37.3850}{474.789} - 0.02386 = 0.05488$$

Thus

$$\frac{R_{o,ins}}{R_{i,ins}} = 1.0564 \quad or \quad R_{i,ins} = \frac{0.04445}{1.0564} = 0.04208$$

Since the outside radius, R_0, of the mild steel pipe is equal to the inside radius, $R_{i,ins}$ of the 85% magnesia insulation, the thickness of magnesia insulation required is equal to ($R_0 - R_i$). That is

$$R_o - R_{i,ins} = 0.04445 - 0.04208 = 0.00237 \ m. \quad Ans$$

Example 4.18: Under what heat transfer conditions is the Poisson's Equation best applied?

Answer: The Poisson's equation represents the steady state situation where heat conduction takes place with internal generation of heat at a constant rate, Q_V, per unit volume. It is stated as

$$\frac{Q_V}{k} + \frac{\partial^2 T}{\partial x^2} + \frac{\partial^2 T}{\partial y^2} + \frac{\partial^2 T}{\partial z^2} = 0 \qquad (4.26)$$

from equation (2.09), or in abbreviated form

$$\frac{Q_V}{k} + \nabla^2 T = 0 \qquad (4.26a)$$

Common areas of application of this equation are in the design or analysis of electrical heating elements, setting of concrete and of heat generation from nuclear fuel rods, etc.

Example 4.19: Derive the expressions for the temperature distribution and the heat flux in an electric heating element in steady state heating in which Q_V is constant. Assume that the element is in the form of a cylindrical wire.

Answer: Because the wire can be regarded as a long solid cylinder, the use of cylindrical co-ordinates makes analysis simpler. Taking the surface temperature as constant, one dimensional analysis, in such co-ordinates, gives

$$\frac{Q_V}{k} + \frac{\partial^2 T}{\partial r^2} + \frac{1}{r}\frac{\partial T}{\partial r} = 0 \qquad (4.27)$$

Because the left hand side of this equation is zero, it can be expressed as a total differential. Thus

$$\frac{Q_V}{k} + \frac{d^2T}{dr^2} + \frac{1}{r}\frac{dT}{dr} = \frac{Q_V}{k} + \frac{1}{r}\frac{d}{dr}\left(r\frac{dT}{dr}\right) = 0 \qquad (4.28)$$

On rearrangement, equation (4.28) becomes

$$\frac{d}{dr}\left(r\frac{dT}{dr}\right) = -r\frac{Q_V}{k} \qquad (4.28a)$$

which, on first integration, becomes

$$r\frac{dT}{dr} = -r^2\frac{Q_V}{2k} + C_1 \qquad (4.28b)$$

where C_1 is a constant of integration. Equation (4.28b) can be further rearranged as

$$\frac{dT}{dr} = -r\frac{Q_V}{2k} + \frac{C_1}{r} \qquad (4.28c)$$

When equation (4.28c) is integrated with the following boundary conditions

$$r = 0, \quad T = T_o, \quad \left.\frac{dT}{dr}\right|_{r=0} = 0 \qquad (4.28d)$$

which show C_1 to be zero, we get

$$T = -r^2\frac{Q_V}{4k} + C_2 \qquad (4.29)$$

where the constant, C_2, is given by

$$C_2 = T_o + \frac{R_o{}^2 Q_V}{4k} \qquad (4.29a)$$

Thus, the temperature distribution is, from equations (4.29) and (4.29a)

$$T = -\frac{Q_V}{4k}r^2 + T_o + \frac{R_o{}^2 Q_V}{4k} = T_o + \frac{R_o{}^2 Q_V}{4k}\left[1 - \left(\frac{r}{R_o}\right)^2\right] \qquad (4.30)$$

Since

$$q_r = -k\frac{\partial T}{\partial r}$$

the heat flux is

$$q_r = -k\frac{\partial T}{\partial r} = -k.-\frac{rQ_V}{2k} = \frac{rQ_V}{2} \qquad (4.31)$$

If $S_R = 2\pi RL$ is the surface area of the cylinder at radius R, at which the heat flux is q_R, the total heat transferred by conduction, Q, in the r-direction is given by

$$Q = q_R.S_R = \frac{RQ_V}{2}.2\pi RL = \pi R^2 LQ_V \qquad (4.32)$$

Example 4.20: Nichrome wire, having a resistivity of 100 µOhm-cm, is to be used as a heating element in a 10 kW heater. The Nichrome wire surface temperature should not exceed 1590 K. Input power is available at 12 volts. Determine the diameter of the Nichrome wire for a 1-meter long heater. You are given that surrounding air temperature is 360K and the outside surface coefficient is 1.14 kW/m²K.

Answer: Let the diameter be d, m

Then for 1 m length,

$$Volume\ of\ wire = \frac{\pi d^2 L}{4} = \frac{\pi d^2}{4}, m^3$$

Rate of heat generation per unit volume, Q_V is

$$Q_V = \frac{10.000}{\pi d^2/4} = \frac{40.000}{\pi d^2}, W/m^3 \qquad (1)$$

Heat transferred from the surface of the wire

$$Q_T = h.\Delta T.A = 1.14\ x\ 1000\ x\ (1590$$
$$- 360)\ x\ \pi d, \frac{kW}{m^2 K}.\frac{W}{kW}.K.m^2$$

81

$$= 1140 \; x \; 1230 \; x \; \pi d, \quad \frac{W}{m^2 K} . K . m^2 \tag{2}$$

But from equation (4.36), $Q_T = \pi R_o^2 L Q_V$.

Hence from equations (2) and (4.36)

$$\pi R_o^2 L Q_V = 1140 \; x \; 1230 \; x \; \pi d \tag{3}$$

Since $R_0 = d/2$ from which

$$\pi R_o^2 L Q_V = \pi \frac{d^2}{4} x 1 x \frac{40{,}000}{\pi d^2} = 1140 \; x \; 1230 \; x \; \pi d$$

$$d = \frac{10{,}000}{\pi \; x \; 1140 \; x \; 1230} = 2.27 \; x \; 10^{-3} \; m = 2.27 \; mm. \quad Ans$$

Example 4.21: Derive the expression for steady state, one dimensional heat conduction in the radial direction in a hollow sphere.

Answer: The Laplace's equation in spherical co-ordinates in the r-direction is

$$\frac{1}{r^2} \frac{\partial}{\partial r} \left(r^2 \frac{\partial T}{\partial r} \right) = 0)$$

This is equivalent to

$$\frac{1}{r^2} \frac{d}{dr} \left(r^2 \frac{dT}{dr} \right) = 0)$$

which means that

$$r^2 \frac{dT}{dr} = C, a \; constant \tag{4.33}$$

It is not difficult to see that the boundary conditions would be $T = T_o$ = at $r = R_o$ and $T = T_i$ at $r = R_i$. Thus

$$\int_{T_i}^{T_o} dT = T_o - T_i = C \int_{R_i}^{R_o} \frac{dr}{r^2} = C \left(\frac{1}{R_i} - \frac{1}{R_o} \right) \tag{4.33a}$$

82

Hence

$$C = \frac{T_o - T_i}{\left(\frac{1}{R_i} - \frac{1}{R_o}\right)} \tag{4.33b}$$

The temperature distribution becomes, from equations (4.33a) and (4.33b)

$$T - T_i = \frac{T_o - T_i}{\left(\frac{1}{R_i} - \frac{1}{R_o}\right)} \left(\frac{1}{R_i} - \frac{1}{r}\right) \tag{4.34}$$

Since

$$q_r = -k\frac{dT}{dr}$$

and

$$\frac{dT}{dr} = \frac{C}{r^2}$$

with equation (4.33b)

$$q_r = \frac{k}{r^2} \frac{T_i - T_o}{\left(\frac{1}{R_i} - \frac{1}{R_o}\right)} \tag{4.35}$$

If $S_r = 4\pi r^2$ is the surface area of the sphere at radius r, at which the heat flux is q_r, the total heat transferred by conduction, Q, in the r-direction is given by

$$Q = q_r S_r = q_r . 4\pi r^2 = 4\pi k \frac{T_i - T_o}{\left(\frac{1}{R_i} - \frac{1}{R_o}\right)} \tag{4.36}$$

Example 4.22: Derive the expression for steady state, one dimensional heat conduction in the radial direction in a hollow sphere when k varies with temperature.

Answer: If $k = k_o [1 + \beta(T - T_o)]$, an analysis, similar to those for the flat plate and cylinder can be done using a mean thermal conductivity, k_m.

Example 4.23: Derive the expression for steady state, one dimensional heat conduction in the radial direction in composite spheres.

Answer: For composite spheres, use same procedure as for composite plates or cylinders

Example 4.24: An asbestos pad, square in cross-section, measures 0.1 m on a side at its large end; the sides decrease linearly to 0.05 m at the small end. The pad is 0.15 m high with the temperatures at the large and small ends held at 700K and 340K respectively. If one dimensional heat conduction is assumed and convective losses from the sides are neglected, what will be the rate of heat conduction through the asbestos?

Answer: This example illustrates the fact that flat plates or walls, cylindrical or spherical objects are not the only configurations which feature in heat transfer by conduction.

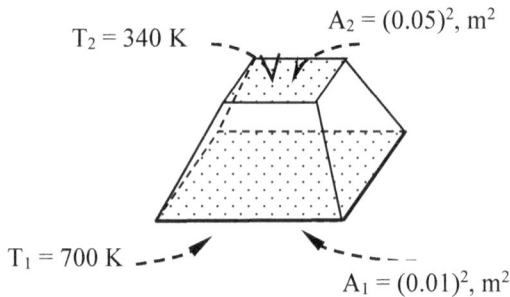

$T_2 = 340$ K $\qquad A_2 = (0.05)^2,\ m^2$

$T_1 = 700$ K $\qquad A_1 = (0.01)^2,\ m^2$

The problem statement is illustrated in the diagram above.

Fourier's first law is

$$q = -k\frac{dT}{dx} = \frac{Q}{A} \qquad (1)$$

84

where Q is total heat transferred and A is area normal to the direction of heat transfer. Since A is a function of x in this system, q is not a constant but $Q = q.A$ is.

If A is area at any x starting from A_1 where $x_1 = 0$, then

$$\frac{A - A_1}{x - 0} = \frac{A_2 - A_1}{0.15 - 0}$$

or with $A_2 = 0.052$ m^2 and $A_1 = 0.102$ m^2,

$$A = A_1 + \frac{x(A_2 - A_1)}{0.15} = (0.1)^2 + \frac{x[(0.05)^2 - (0.1)^2]}{0.15}$$

$$= 0.01 - 0.05x \tag{2}$$

Equation (2) gives the cross-sectional area of the asbestos pad as a function of x. We can, now, obtain an expression for Q as a function of T and x as follows

$$Q = q.A = -k\frac{dT}{dx}(0.01 - 0.05x) \tag{3}$$

Rearranging equation (3) with $T_2 = 340$K and $T_1 = 700$K,

$$\int_0^{0.15} \frac{dx}{(0.01 - 0.05x)} = -\frac{k}{Q}\int_{700}^{340} dT$$

and integrating we get

$$-\frac{1}{0.05}\ln\left(\frac{0.01 - 0.05 \times 0.15}{0.01}\right) = \frac{k}{Q}(700 - 340) \tag{4}$$

Since $k = 0.214$ at the mean temperature between 700 K and 340 K, that is at 520 K, equation (4) reduces to

$$27.7259 = \frac{0.214 \times 360}{Q}$$

From which Q = 2.7786 W, Ans.

References For Chapter Four

1.   Carslaw, H.S and Jaeger, J.C., *Conduction of Heat in Solids*, Clarendon Press, Oxford, UK, 1959

2.   Coulson J.M. and Richardson J.F. *Chemical Engineering Vol. 1*, Pergamon Press, Oxford, UK, 1978

3.   Heldman D.R: *Food Process Engineering*, Chapter 3, AVI Publishing Co; Westport, Connecticut, USA. 1975

4.   Welty J.R: *Engineering Heat Transfer*, John Wiley and Sons, New York, USA. 1978

5.   Welty J.R; Wicks C.E and Wilson R.E: *Fundamentals of Momentum, Heat and Mass Transfer*; 2nd Edition; John Wiley and Sons, New York, USA, 1976.

CHAPTER FIVE
MORE STEADY STATE ONE DIMENSIONAL HEAT CONDUCTION

Example 5.01:   In what situations would solution methods different from those in Chapter Four be necessary?

Answer: These would be situations in which the methods of Chapter Four are not applicable or satisfactory. Such situations may have to do with the nature of the material involved, the mathematical complexity which may arise with using these methods, or when other modes of heat transfer, such as convection or radiation, are involved, etc.

Example 5.02: What are these other methods?

Answer: These other methods include

- Graphical methods which, historically, went out of date once the use of digital computers became widespread
- Numerical methods which are more important for multidimensional heat conduction, and
- Thermal resistance circuit methods in which an analogy is developed between heat transfer and electrical conduction. They are, particularly, useful in the analysis of complex configurations.

Only the numerical and electrical circuit methods will be illustrated.

Example 5.03: Outline the more common numerical methods in use for the solution of steady state, one dimensional heat conduction.

Answer: There are quite a few numerical methods for solving ordinary differential equations depending on whether the problem is an initial value or a boundary value problem. Most are based on the valid assumption that the terms in a Taylor series expansion may be used to represent, in difference form, the derivatives in the differential equation to be solved.

Thus, using only finite differences (forward, central or backward differences) and an iteration technique, differential equations which, otherwise, would demand tedious and complex analytical procedures, become relatively easy to solve.

The Taylor series expansion in temperature, T, around a point, x_i, on increment of h, is given by

$$T(x_i + h) = T(x_i) + h\left(\frac{dT}{dx}\right)_i + \frac{h^2}{2!}\left(\frac{d^2T}{dx^2}\right)_i + \frac{h^3}{3!}\left(\frac{d^3T}{dx^3}\right)_i + .. \, (5.01)$$

When it is expanded in temperature, T, around the same point x_i, but on decrement of h, it is given by

$$T(x_i - h) = T(x_i) - h\left(\frac{dT}{dx}\right)_i + \frac{h^2}{2!}\left(\frac{d^2T}{dx^2}\right)_i - \frac{h^3}{3!}\left(\frac{d^3T}{dx^3}\right)_i + .. \, (5.02)$$

Neglecting terms $\frac{d^3T}{dx^3}$ and higher and adding the two equations gives

$$\left(\frac{d^2T}{dx^2}\right)_i = \frac{T(x_i + h) + T(x_i - h) - 2T(x_i)}{h^2} \qquad (5.03)$$

with an error of the order h^3. Neglecting the terms $\frac{d^2T}{dx^2}$ and higher from equation (5.01) gives

$$\left(\frac{dT}{dx}\right)_i = \frac{T(x_i + h) - T(x_i)}{h} \qquad (5.04)$$

with an error of the order h^2. Similarly if the terms $\frac{d^2T}{dx^2}$ and higher are neglected in equation (5.02)

$$\left(\frac{dT}{dx}\right)_i = \frac{T(x_i) - T(x_i - h)}{h} \tag{5.05}$$

If equation (5.02) is subtracted from equation (5.01) and terms $\frac{d^3T}{dx^3}$ and higher are neglected we get

$$\left(\frac{dT}{dx}\right)_i = \frac{T(x_i + h) - T(x_i - h)}{2h} \tag{5.06}$$

If it is realised that

$$h = \Delta x = (x_i + h) - x_i \tag{5.07}$$

Then equation (5.04), expressed as

$$\left(\frac{dT}{dx}\right)_i = \frac{T(x_i + h) - T(x_i)}{\Delta x} \tag{5.08}$$

is the first forward difference representation of $\frac{dT}{dx}$ and equation (5.05) expressed as

$$\left(\frac{dT}{dx}\right)_i = \frac{T(x_i) - T(x_i - h)}{\Delta x} \tag{5.09}$$

is the first backward difference representation of $\frac{dT}{dx}$. Equation (5.06), expressed as

$$\left(\frac{dT}{dx}\right)_i = \frac{T(x_i + h) - T(x_i - h)}{2\Delta x} \tag{5.10}$$

is the central difference representation of $\frac{dT}{dx}$. Armed with these, therefore, any steady state, first and second order differential equation may be solved numerically by an appropriate substitution of these difference formulae.

Take, for example, the Laplace's equation, in the x-direction.

$$\frac{d^2T}{dx^2} = 0$$

Using the first forward difference, equation (5.03)

$$\left(\frac{d^2T}{dx^2}\right)_i = \frac{T(x_i + h) + T(x_i - h) - 2T(x_i)}{h^2} \qquad (5.11)$$

It can be seen that

$$T(x_i + h) + T(x_i - h) - 2T(x_i) = 0 \qquad (5.12)$$

If the limits on x are a and b and i represents 0, 1, 2, 3, 4, etc. up to n values that we wish to use, then h is given by

$$h = \frac{b - a}{n} \qquad (5.13)$$

and $\quad x_1 = a + h, \ x_2 = a + 2h, \ ... \ ... \ x_n = a + (n-1)h \qquad (5.14)$

From the boundary conditions,

$$x_o = a, \qquad x_n = b, \qquad T_o \ is \ known, \ T_n \ is \ known \qquad (5.15)$$

If we chose n = 4, then, from equation (5.12)

$$-2T_1 + T_2 + 0 = -T_o$$
$$T_1 - 2T_2 + T_3 = 0$$
$$0 + T_2 - 2T_3 = -T_n$$

This can be solved as three simultaneous equations or as a matrix in three unknowns

$$\begin{pmatrix} -2 & 1 & 0 \\ 1 & -2 & 1 \\ 0 & 1 & -2 \end{pmatrix} \begin{pmatrix} T_1 \\ T_2 \\ T_3 \end{pmatrix} = \begin{pmatrix} -T_o \\ 0 \\ -T_n \end{pmatrix} \qquad (5.16)$$

Although the above example appears, relatively, straightforward, numerical methods can get complex. They are regarded as explicit, if the unknown quantity is solved for directly, or implicit if the unknowns have to be solved for simultaneously, as above, or mixed (weighted average) if both the explicit and implicit methods are both employed usually with a weighting factor, F, whose value is between 0 and 1.

The common mathematical procedures for explicit solutions of initial value problems are the Newton's iteration and Runge-Kutta methods in single step iteration and the Adams-Bashforth/Adams-Moulton methods in multi-step procedures.

Example 5.03: What is a thermal resistance circuit?

Answer: A thermal resistance circuit represents the resistance to heat flow as if it were an electric resistance. Thus, in analogy to electrical circuits, the heat transferred is analogous to electric current, temperature difference is analogous to voltage difference and thermal resistance is analogous to electrical resistance. These thermal analogies are summarised in Table 5.1.for the three modes of heat transfer.

Thus if, for heat conduction, $R = \dfrac{L}{kA}$ and $\Delta T = T_1 - T_2$, then

$$Q = \frac{T_1 - T_2}{\dfrac{L}{kA}} = \frac{\Delta T}{R} \qquad (5.17)$$

Table 5.1: Thermal Circuit Definitions of Thermal Resistance (Wikipedia, 2009)

Mode of Heat Transfer	Quantity of Heat Transferred	Thermal Resistance
Conduction	$Q = \dfrac{T_1 - T_2}{\dfrac{L}{kA}}$	$\dfrac{L}{kA}$
Convection	$Q = \dfrac{T_1 - T_2}{\dfrac{1}{h_{convection} \cdot A_1}}$	$\dfrac{1}{h_{convection} \cdot A_1}$
Radiation	$Q = \dfrac{T_1 - T_2}{\dfrac{1}{h_{radiation} \cdot A_1}}$	$\dfrac{1}{h_{radiation} \cdot A_1}$

Example 5.04: What is thermal capacitance in heat conduction?

Answer: Thermal capacitance, C, is defined as the heat energy flow per unit mass per unit temperature difference. A closer look at this definition suggests that it is, in fact, the definition of specific heat. Thus

$$C = \frac{Q}{m\Delta T} \tag{5.18}$$

where m is mass.

Example 5.05: How would the thermal resistance circuit analysis be used in the case of heat transfer through composite parallel walls?

Answer: The thermal situation may be represented as a thermal circuit as shown in the diagram below. Since the resistances are in series, total resistance $R_T = R_1 + R_2$ and

$$Q = \frac{\Delta T}{R_T} = \frac{T_1 - T_2}{R_1 + R_2} = \frac{T_1 - T_2}{\dfrac{L_1}{k_1 A} + \dfrac{L_2}{k_2 A}} \qquad (5.19)$$

This can be extended to any number of parallel walls.

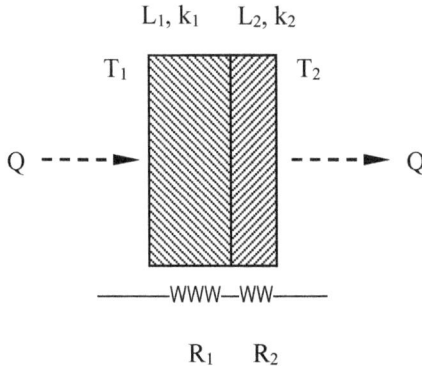

Other situations arise in which the thermal resistances, just like in electrical circuits, are not in parallel. A typical case is that in which a bolt or fastener of a different material is used to hold the wall in place, as illustrated below. In such a case, thermal resistance occurs in parallel and total resistance is

$$\frac{1}{R_T} = \frac{1}{R_1} + \frac{1}{R_2} \qquad (5.20)$$

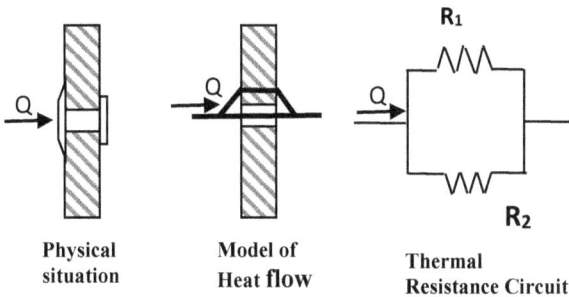

Physical situation   Model of Heat flow   Thermal Resistance Circuit

This gives Q as

$$Q = \frac{\Delta T}{R_T} = \frac{(T_1 - T_2)(R_1 + R_2)}{R_1 R_2} \tag{5.21}$$

Remember that R, for each material, is, still, equal to its L/kA.

Example 5.06: Fins are, normally, associated with fish. What are fins in heat transfer?

Answer: Fins are extended surfaces to regular heat transfer equipment surfaces for the purpose of improving heat transfer efficiency and reducing overall equipment size and operating cost. They are popular in heat exchangers, in chemical reactors and in other heat transfer equipment in which efficient heat transfer is desired,. In most cases in which fins are used, conduction is not the only mode of heat transfer. Convection and radiation are also, generally, involved.

Example 5.07: Derive the expressions for steady state heat transfer from fins and extended surfaces

Answer: In order not to make the analysis unduly complicated at this stage, let us treat only the case in which convection is the only other mode of heat transfer involved. We shall concentrate on deriving general expressions for the temperature distribution and the heat flux during heat transfer from an extended surface such as a fin of constant cross - section.

The General Case of a Fin of General Cross –section

Consider a fin of general section attached to a mother surface as shown below. A heat balance on element of thickness, Δx, of the fin, gives

$Heat\ in + Heat\ generated$
$\qquad\qquad = Heat\ out + Heat\ accummulated$ (5.22)

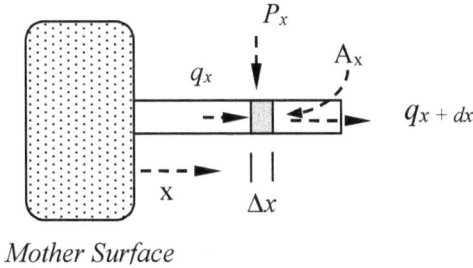

Mother Surface

At steady state, and assuming no heat generation

$$Heat\ in = Heat\ out \qquad (5.22a)$$

That is

$$q_x A_x = q_x A_x + A_x \frac{\partial q_x}{\partial x} \Delta x + h P_x (T - T_\infty) \Delta x \qquad (5.22b)$$

where h_x = heat transfer coefficient at the surface of the element A_x,

P_x = perimeter of this element at x,
T = temperature of the element,
T_∞ = temperature of surrounding fluid

Equation (5.22b) reduces, on simplification, to

$$-\frac{\partial A_x q_x}{\partial x} = h P_x (T - T_\infty) \qquad (5.22c)$$

Since $q = -k\frac{\partial T}{\partial x}$, generally,

$$-\frac{\partial \left(-k\frac{\partial T}{\partial x} A_x\right)}{\partial x} = h P_x (T - T_\infty) \qquad (5.22d)$$

For a uniform cross-section, A_x = constant = A. Equation (5.22d) becomes

$$kA\frac{d^2 T}{dx^2} - h P_x (T - T_\infty) = 0$$

or

$$\frac{d^2T}{dx^2} - \frac{hP_x}{kA}(T - T_\infty) = 0 \tag{5.23}$$

By letting $\theta = T - T_\infty$ and $m = \frac{hP_x}{kA}$ and since $d\theta = dT$, equation (5.23) becomes

$$\frac{d^2\theta}{dx^2} - m\theta = 0 \tag{5.23a}$$

This has the solution,

$$\theta = C_1 e^{mx} + C_2 e^{-mx} \tag{5.24}$$

or

$$\theta = A_1 \cosh mx + B_1 \sinh mx \tag{5.24a}$$

which is the temperature distribution, and C_1, C_2, A_1, B_1 are constants determined by the boundary conditions. The heat flux, q, is then

$$q = -k\frac{dT}{dx}\bigg|_{x=0} = -k\frac{d\theta}{dx}\bigg|_{x=0} \tag{5.25}$$

If equation (5.24) is used

$$q = -k|mC_1 e^{mx} - mC_2 e^{-mx}|_{x=0} = -k(C_1 - C_2)m \tag{5.26}$$

If equation (5.24a) is used

$$q = -k| A_1 m \sinh mx + B_1 m\cosh mx|_{x=0} = -kmB_1 \tag{5.27}$$

a   Specific Cases of Fins of Constant Cross -section

Let us use the solutions based on equation (5.24) for these cases.

Case I: Fin is very long

This is equivalent to the boundary conditions

$$x = 0 \qquad \theta = \theta_o \qquad\qquad (5.28)$$
$$x = \infty \qquad \theta = 0 \qquad\qquad (5.28a)$$

so that at

$$x = 0 \qquad \theta_o = C_1 + C_2 \qquad\qquad (5.28b)$$

$$x = \infty \qquad 0 = C_1 \, x \, \infty + C_2 \, x \, 0 \qquad\qquad (5.28c)$$

Hence
$$C_1 = 0, \quad C_2 = \theta_o \qquad\qquad (5.28d)$$

and
$$\theta = \theta_o e^{-mx} \qquad\qquad (5.29)$$

That is, the temperature distribution is

$$T = T_\infty + (T_o - T_\infty)e^{-\frac{hP_x}{kA}x} \qquad\qquad (5.29a)$$

The heat flux is then

$$q = -k \frac{d\theta}{dx}\bigg|_{x=0} = -k \cdot -m\theta_o|e^{-mx}|_{x=0} = k \cdot m \cdot \theta_o \qquad (5.29b)$$

That is

$$q = \frac{hP_x}{A}(T_o - T_\infty) \qquad\qquad (5.29c)$$

Case II: Fin is of finite length, L

The boundary conditions become, in this case

$$x = 0 \qquad \theta = \theta_o \qquad\qquad (5.30)$$
$$x = L \qquad \theta = \theta_L \qquad\qquad (5.30a)$$

so that at
$$x = 0 \qquad \theta_o = C_1 + C_2 \qquad\qquad (5.30b)$$
$$x = L \qquad \theta_L = C_1 e^{mL} + C_2 e^{-mL} \qquad\qquad (5.30c)$$

Multiplying (5.30b) by e^{mL}

97

$$\theta_o e^{mL} = C_1 e^{mL} + C_2 e^{mL} \qquad (5.30d)$$

Subtracting equation (2.62c) from equation (2.62d) and re-arranging

$$C_2 = \frac{\theta_o e^{mL} - \theta_L}{e^{mL} - e^{-mL}} \qquad (5.30e)$$

Multiplying (5.30b) by e^{-mL}

$$\theta_o e^{-mL} = C_1 e^{-mL} + C_2 e^{-mL} \qquad (5.30f)$$

Subtracting equation (5.30f) from equation (5.30c) and rearranging

$$C_1 = \frac{\theta_L - \theta_o e^{-mL}}{e^{mL} - e^{-mL}} \qquad (5.30g)$$

Substituting equations (5.30e) and (5.30g) into equation (5.24)

$$\theta = \frac{\theta_L - \theta_o e^{-mL}}{e^{mL} - e^{-mL}} e^{mx} + \frac{\theta_o e^{mL} - \theta_L}{e^{mL} - e^{-mL}} e^{-mx} \qquad (5.31)$$

This is the temperature distribution. This can be simplified as

$$\theta = \frac{\theta_L - \theta_o e^{-mL}}{e^{mL} - e^{-mL}} e^{mx} + \frac{\theta_o e^{mL} - \theta_L + \theta_o e^{-mL} - \theta_o e^{-mL}}{e^{mL} - e^{-mL}} e^{-mx}$$

to get

$$\theta = \frac{e^{mx} - e^{mx}}{e^{mL} - e^{-mL}} (\theta_L - \theta_o e^{-mL}) + \theta_o e^{-mx} \qquad (5.32)$$

which shows that as $L \to \infty$, equation (5.32) reduces to equation (5.29), the case of a very long fin. For the fin of finite length, however, the heat flux is then

$$q = -k \frac{d\theta}{dx}\Big|_{x=0} = -k \left| (me^{mx} + me^{-mx}) \left(\frac{\theta_L - \theta_o e^{-mL}}{e^{mL} - e^{-mL}} \right) - m\theta_o e^{-mx} \right|_{x=0}$$

That is

$$q = k \cdot m \left[\theta_o - 2 \left(\frac{\theta_L - \theta_o e^{-mL}}{e^{mL} - e^{-mL}} \right) \right] \qquad (5.33)$$

Case III: Fin is insulated at the end

The boundary conditions become, in this case

$$x = 0 \qquad \theta = \theta_o \qquad\qquad\qquad (5.34)$$

$$x = L \qquad \left.\frac{d\theta}{dx}\right|_{x=L} = 0 \qquad\qquad (5.34a)$$

so that at

$$x = 0 \qquad \theta_o = C_1 + C_2 \qquad\qquad (5.34b)$$

$$x = L \qquad C_1 e^{mL} - C_2 e^{-mL} = 0 \qquad (5.34c)$$

Multiplying (5.34b) by e^{mL}

$$\theta_o e^{mL} = C_1 e^{mL} + C_2 e^{mL} \qquad (5.34d)$$

Subtracting   equation (5.34c) from equation (5.34d) and re-arranging

$$C_2 = \frac{\theta_o e^{mL}}{e^{mL} + e^{-mL}} \qquad (5.34e)$$

Multiplying (5.34b) by e^{-mL}

$$\theta_o e^{-mL} = C_1 e^{-mL} + C_2 e^{-mL} \qquad (5.34f)$$

Adding equation (5.34c) to equation (5.34f) and rearranging

$$C_1 = \frac{\theta_o e^{-mL}}{e^{mL} + e^{-mL}} \qquad (5.34g)$$

Substituting equations (5.34e) and (5.34g) into equation (5.24)

$$\theta = \frac{\theta_o e^{-mL}}{e^{mL} + e^{-mL}} e^{mx} + \frac{\theta_o e^{mL}}{e^{mL} + e^{-mL}} e^{-mx}$$

which simplifies to

$$\theta = \frac{\theta_o[e^{m(L-x)} + e^{-m(L-x)}]}{e^{mL} + e^{-mL}} = \frac{\theta_o \cosh m(L-x)}{\cosh mL} \tag{5.35}$$

The heat flux is then

$$q = -k\frac{d\theta}{dx}\Big|_{x=0} = +k\left|\frac{\theta_o m. \sinh m(L-x)}{\cosh mL}\right|_{x=0}$$

$$= k.m.\theta_o \tanh(mL) \tag{5.35a}$$

Case IV: Conduction to, is equal to conduction from, the end of the Fin

The boundary conditions are

$$x = 0 \qquad \theta = \theta_o \tag{5.36}$$

$$x = L \qquad -k\frac{d\theta}{dx}\Big|_{x=L} = h\theta \tag{5.36a}$$

so that at

$$x = 0 \qquad \theta_o = C_1 + C_2 \tag{5.36b}$$

$$x = L \qquad -k(mC_1 e^{mL} - me^{-mL})$$

$$= h(C_1 e^{mL} + C_2 e^{-mL})$$

That is

$$C_1(h + mk)e^{mL} + C_2(h - mk)e^{-mL} = 0 \tag{5.36c}$$

Multiplying (5.36b) by $(h + mk)e^{mL}$

$$C_1(h + mk)e^{mL} + C_2(h + mk)e^{mL} = \theta_o(h + mk)e^{mL} \tag{5.36d}$$

Subtracting equation (5.36c) from equation (5.36d) and rearranging

$$C_2 = \frac{\theta_o(h + mk)e^{mL}}{h(e^{mL} - e^{-mL}) + mk(e^{mL} + e^{-mL})} \tag{5.36e}$$

100

Multiplying (5.36b) by $(h - mk)e^{-mL}$

$$C_1(h - mk)e^{-mL} + C_2(h - mk)e^{-mL} = \theta_o(h - mk)e^{-mL} \quad (5.36f)$$

Subtracting equation (5.36c) from equation (5.36f) and rearranging

$$C_1 = -\frac{\theta_o(h - mk)e^{-mL}}{h(e^{mL} - e^{-mL}) + mk(e^{mL} + e^{-mL})} \quad (5.36g)$$

Substituting equations (5.36e) and (5.36g) into equation (5.24)

$$\theta = -\frac{\theta_o(h - mk)e^{-mL}}{h(e^{mL} - e^{-mL}) + mk(e^{mL} + e^{-mL})} e^{mx}$$

$$+ \frac{\theta_o(h + mk)e^{mL}}{h(e^{mL} - e^{-mL}) + mk(e^{mL} + e^{-mL})} e^{-mx}$$

This can be simplified to

$$\theta = \theta_o \left[\frac{(e^{m(L-x)} + e^{-m(L-x)}) + (h/mk)(e^{m(L-x)} - e^{-m(L-x)})}{(e^{mL} + e^{-mL}) + (h/mk)(e^{mL} - e^{-mL})} \right] \quad (5.37)$$

or to

$$\theta = \theta_o \left[\frac{\cosh m(L - x) + (h/mk)\sinh m(L - x)}{\cosh mL + (h/mk)\sinh mL} \right] \quad (5.37a)$$

The heat flux is then

$$q = -k\frac{d\theta}{dx}\bigg|_{x=0}$$

$$= +k\left|\theta_o\left\{\frac{m.\sinh m(L - x) + (h/mk)m\cosh m(L - x)}{\cosh mL + (h/mk)\sinh mL}\right\}\right|_{x=0}$$

That is

$$q = mk\theta_o \left\{ \frac{\sinh mL + (h/mk)\cosh mL}{\cosh mL + (h/mk)\sinh mL} \right\} \qquad (5.38)$$

Example 5.08: What is meant by the effectiveness of a fin

Answer: The idea of effectiveness of a fin arises from the fact that the temperature at any point along the length of the fin, $T(x)$, will, in practice, be less than the temperature at the root or mother surface, T_0. Thus, for $x > 0$, $T(x) < T_0$. This means that the actual heat transferred through the fins, Q_{actual}, will always be less than the maximum heat transferable, Q_{max}, if all of the fin was at the same temperature as the root of the fin or the mother surface. The fin effectiveness, ε, is, thus, defined as

$$\varepsilon = \frac{Q_{actual}}{Q_{max}} \qquad (5.39)$$

Example 5.09: Outline the derivation of the expression for the actual heat transferred in a fin.

Answer: The maximum heat transferred from a fin surface, assuming all of the fin is at the same temperature as the mother surface and losing heat, by convection only, to an environment which is at the temperature, T_∞, is given by

$$Q_{max} = hA_f(T_o - T_\infty) \qquad (5.40)$$

where h is the heat transfer coefficient, based on the primary surface, for heat convection from the surface of the fin to the surroundings, A_f is the surface area of the fin through which heat transfer takes place and T_∞ is the temperature of the surroundings. Using equations (5.39) and (5.40), the actual heat transferred is seen to be

$$Q_{actual} = \varepsilon. Q_{max} = \varepsilon. hA_f(T_o - T_\infty) \qquad (5.41)$$

This definition, in equation (5.41), is, especially, useful for complex shapes for which Q_{actual} is difficult to obtain analytically. It is usual industry practice to estimate ε from charts in which some characteristic parameter of the fin is plotted against its effectiveness, ε.

Example 5.10: Outline the derivation of the expression for the total heat transferred into or from a finned surface.

Answer: The total heat transferred, Q_T, to or from a finned surface, is the sum of the heat transferred through the primary or mother surface and that transferred through the fins. That is

$$Q_T = Q_{primary\ surface} + Q_{fins} = hA_f(T_o - T_\infty) + \varepsilon. hA_f(T_o - T_\infty)$$

$$= h(A_o + \varepsilon. A_f)(T_o - T_\infty) \qquad (5.42)$$

A_0 is the area of the primary surface.

In theory h is not the same at the mother surface as at the fin surface but is assumed to be so for lack of accurate data.

Example 5.11: It is desired to estimate the total heat flow from a stainless steel cylinder 2.5 m in diameter and 1.0 m high and maintained at 353 K. The outside surface of this cylinder has, attached to it, fins in the form of cylindrical pins 1.9 cm in diameter, 25.4 mm long, located at 25 4 mm centres and in contact with air at 302 K.

If the heat transfer coefficient on the air side is 15.3 W/m²K and the fin effectiveness on this side is 0.97, determine the heat transferred on the air side.

Answer: The heat transferred, Q, is given by

$$Q_T = h(A_o + \varepsilon A_f)(T_o - T_\infty) \qquad (5.42)$$

where h = heat transfer coefficient = 15.3 W/m²K, A_0 = area of un-finned surface, A_f = area of finned surface, ε = fin effectiveness = 0.97, T_0 = temperature of un-finned surface = 353 K, T_∞ = temperature of surroundings = 302 K

<u>To estimate A_o and A_f</u>

Number of fins along cylinder height $= \frac{100 \, cm}{2.54 \, cm} = 39.37 \approx 40$

Number of fins along circumference of cylinder $= \frac{\pi \, x \, 2.5}{0.0254} = 309$

Total Number of fins in cylinder = 309 x 40 = 12360

Surface Area of each fin

$$= \frac{\pi d^2}{4} + \pi dL = \pi \left[\frac{(0.019)^2}{4} + 0.019 \, x \, 0.0254 \right] = 0.0018 \, m^2$$

Total fin area

$$A_f = 12360 \, x \, 0.0018 = 22.248 \, m^2$$

Total un-finned area = A_0 = total cylinder surface area - total cross- section area of fins

$$= \pi \, x \, 2.5 \, x \, 1 - 12360 \, x \, \frac{\pi}{4} \, x \, (0.019)^2 = 4.350 \, m^2$$

Q_T is given by

$$Q_T = 15.3(4.350 + 0.97 \, x \, 22.248)(353 - 302)$$
$$= 20233.6 \, W. \; Ans$$

References For Chapter Five

1. Carslaw, H.S and Jaeger, J.C., *Conduction of Heat in Solids*, Clarendon Press, Oxford, UK, 1959

2. Coulson J.M. and Richardson J.F. *Chemical Engineering Vol. 1*, Pergamon Press, Oxford, UK, 1978

3. Heldman D.R: *Food Process Engineering*, Chapter 3, AVI Publishing Co; Westport, Connecticut, USA. 1975

4. Welty J.R: *Engineering Heat Transfer*, John Wiley and Sons, New York, USA. 1978

5. Welty J.R; Wicks C.E and Wilson R.E: *Fundamentals of Momentum, Heat and Mass Transfer*; 2nd Edition; John Wiley and Sons, New York, USA, 1976.

CHAPTER SIX
STEADY STATE, MULTI-DIMENSIONAL, HEAT CONDUCTION

Example 6.01: What is the rationale for the search for solutions for multi-dimensional heat conduction in isotropic solids?

Answer: Although one dimensional solutions of the heat conduction equations are much more popular, certain situations arise where two or three dimensional solutions are necessary. Such cases arise in metallurgical processes, food processing, electronic component processing, etc., where non-uniform temperature distribution or heat flux may affect product quality adversely. In some other situations, the contour of the product may be such that non-uniform but known temperature profiles and fluxes will be required to attain a given product quality or processing efficiency.

Example 6.02: What are the common methods employed in the analysis of multi-dimensional heat conduction?

Answer: Four common methods are, generally, employed, depending on the situation. These methods may be classified as analytical, graphical, integral analysis and numerical solution procedures. The analytical, integral and numerical approaches are the preferred methods in organisations with strong mathematical cultures such as research, development and in some cases, design departments. Graphical procedures are much more common in manufacturing and plant operation environments where quick answers, to a practical and fair degree of accuracy, are required and in design departments where complex relationships have been simplified and presented on graphs and charts.

Example 6.03: What do we know about analytical solutions?

Answer: Analytical solutions are, usually, by choice or habit, the first option because they are exact analytic solutions of a given differential equation of heat conduction and can be obtained using standard mathematical procedures. These solutions, usually, have to satisfy some given initial and/or boundary conditions.

Example 6.04: Name the most popular analytical procedures for heat conduction

Answer: The three most popular analytical procedures for heat conduction are the methods of

a. Separation of variables
b. Mathematical Transform methods
c. Product, and superposition, solutions

Example 6.05: Without going into details, what types of equations are encountered in analytical solutions?

Answer: The differential equations encountered in heat conduction are of two main types (there are three) of linear differential equations namely, the linear differential equations classified as **parabolic** such as the heat equation (Fourier's second law of heat conduction), and those classified as **elliptic** such as the Laplace's equation. The third type of linear differential equations is the type classified as **hyperbolic** such as the wave equation (See Table 6.1). The mathematical solutions of these equations are, usually, subject to some initial and boundary conditions associated with the problem in hand.

Table 6.1: Types of Partial Differential Equations

Parabolic	Fourier's second Law of heat conduction	$\dfrac{\partial T}{\partial t} = \alpha \left(\dfrac{\partial^2 T}{\partial x^2} + \dfrac{\partial^2 T}{\partial y^2} + \dfrac{\partial^2 T}{\partial z^2} \right)$
Elliptic	Laplace's equation	$\dfrac{\partial^2 T}{\partial x^2} + \dfrac{\partial^2 T}{\partial y^2} + \dfrac{\partial^2 T}{\partial z^2} = 0$
Hyperbolic	The wave equation	$\dfrac{\partial^2 U}{\partial t^2} = c^2 \left(\dfrac{\partial^2 U}{\partial x^2} + \dfrac{\partial^2 U}{\partial y^2} + \dfrac{\partial^2 U}{\partial z^2} \right)$

Example 6.06: What are initial and boundary conditions in the mathematical solution of heat conduction equations?

Answer: Initial conditions refer to the values of the variables at the start of the process, that is when time, t, is zero. Boundary conditions refer to the values of the variables at the physical boundaries of the system. Usually, one or more of the three known boundary conditions are encountered.

The first type is known as the **Dirichlet** boundary condition. Here, the dependent variable, such as temperature, has a uniform value along the boundary, such as the boundary defined by $x = a$ to $x = b$. The second type is known as the **Neumann** condition. Here, the rate of change of the dependent variable with respect to a given direction is zero while in the third type of boundary condition, known as the **Robin** condition (Zill & Cullen, 1997), this change is constant. These are summarised in Table 6.2 below.

Table 6.2: Types of Boundary Conditions (Zill & Cullen, 1997)

Type 1	Dirichlet condition	$T(L, t) = T_L$, constant
Type 2	Neumann condition	$\left. \dfrac{\partial T}{\partial x} \right|_{x=L} = 0$

Type 3   Robin condition $\dfrac{\partial T}{\partial x}\Big|_{x=L} = -h[T(L,t) - T_\infty]$

Example 6.07: What do we know about graphical solutions?

Answer: Graphical solutions are of two main types. In the first kind, solutions of the heat conduction equations are obtained through a step by step graphical procedure. Typical examples are the flux plotting, the Schmidt plotting and the alternating direction procedures. These methods were popular during the age of the slide rule and log tables but have, more or less, died out in this age of computers because they are tedious, not very accurate and time consuming. They are not treated here but have been discussed in Welty, (1978).

In the second method, also gradually going out of use, final results for a number of typical situations are presented in chart or graphical form in such a way that they are easier and quicker to use than going through the mathematical or graphical solution. These solutions may have been obtained from any of either the analytical, experimental, integral or numerical method. It is common to present the information, on the Y-X axes of a graph, for example, as parameters consisting of a combination of variables with a third or more parameters demarcating regions or lines of similarities of data. Where possible, these parameters are dimensionless

Example 6.08: What do we know about integral solutions?

Answer: This is a successful, analytical, method used in boundary layer analysis in fluid mechanics and in convection/conduction heat transfer (Schlichting, 1969; Welty, Wicks and Wilson, 1976). In the method, the heat flux, within a selected boundary, is obtained by the integration of the heat

110

capacity equation over a control volume. To do this, a temperature profile is assumed within this boundary. Integral analyses are popular with vector configurations and methods.

Example 6.09: What do we know about numerical solutions?

Answer: Numerical solutions are popular methods of solution, mainly because of the availability of computing power, especially in situations where the complexity of the physical problem or the tedium of computation makes other methods inaccurate, inefficient or uneconomic. A numerical solution starts, usually, from some finite difference approximation of the heat conduction equation. This finite difference approximation may be based on either the forward, the backward or central differences.

Numerical solutions are regarded as explicit, if the unknown quantity is solved for directly, implicit if the unknowns have to be solved for simultaneously, or mixed (weighted average) if both the explicit and implicit methods are both employed, often with a weighting factor, F, having values between 0 and 1.
The common mathematical procedures, employed in numerical solutions, are the Newton's or Newton-Raphson iteration procedures, the Runge-Kutta iteration methods using single step iteration and the Adams-Bashforth/Adams-Moulton methods in multi-step procedures (Zill & Cullen, 1997).

When the formulation of a numerical solution is in matrix form, the Gauss - Jordan elimination (pivotal condensation) or the Gauss -Seidel iteration procedures are used because they are efficient for these matrix type solutions.

The choice of which of these methods to use depends, also, on whether the problem is an initial value or boundary value problem, or on whether it is an implicit, explicit or mixed, problem.

Example 6.10: What are the main problems encountered in numerical solutions?

Answer: Numerical solutions have problems of stability, accuracy and convergence. These are better discussed in textbooks on the subject but will be highlighted in the simple examples, in this book, of the use of numerical methods for the solution of multi-dimensional heat conduction equations.

Example 6.11: Outline, with an example, the analytical, multi-dimensional, steady state, solution of heat conduction equations in isotropic solids by the method of separation of variables

Answer: Separation of variables is one of several methods for solving ordinary and partial differential equations, in which the equation is rewritten so that each of the dependent variables occurs on a different side of the equation with its own independent variable. Thus as shown below, using the Laplace's equation, the temperature T, though a simultaneous function of space co-ordinates, can be, by separation of variables technique, broken into terms which are dependent only on one space co- ordinate at a time

Solving the Laplace's Equation in Two Dimensions

(a) *Cartesian Co-ordinates*
The process can be illustrated with a two dimensional example of the Laplace's equation given, in Cartesian co-ordinates as

$$\frac{\partial^2 T}{\partial x^2} + \frac{\partial^2 T}{\partial y^2} = 0 \qquad (6.1)$$

By separation of variables

$$T = T_X(x).T_Y(y) \tag{6.1a}$$

where $T_X(x)$ is a function of x only and independent of y while $T_Y(y)$ is a function of y only and independent of x. Equation (6.1a) can be differentiated to various orders as required such as

$$\frac{\partial T}{\partial x} = T_Y(y)\frac{\partial T_X(x)}{\partial x} \tag{6.1b}$$

$$\frac{\partial^2 T}{\partial x^2} = T_Y(y)\frac{\partial^2 T_X(x)}{\partial x^2} \tag{6.1c}$$

Similarly

$$\frac{\partial T}{\partial y} = T_X(x)\frac{\partial T_Y(y)}{\partial y} \tag{6.1d}$$

$$\frac{\partial^2 T}{\partial y^2} = T_X(x)\frac{\partial^2 T_Y(y)}{\partial y^2} \tag{6.1e}$$

When they are substituted into equation (6.1), we get

$$T_Y(y)\frac{\partial^2 T_X(x)}{\partial x^2} + T_X(x)\frac{\partial^2 T_Y(y)}{\partial y^2} = 0$$

That is

$$\frac{1}{T_X(x)}\frac{\partial^2 T_X(x)}{\partial x^2} = -\frac{1}{T_Y(y)}\frac{\partial^2 T_Y(y)}{\partial y^2} \tag{6.2}$$

Thus, the original equation has been transformed such that the left hand side and the right hand side of equation (6.2) are functions of one variable only. The solution is obtained by setting

$$\frac{1}{T_X(x)}\frac{\partial^2 T_X(x)}{\partial x^2} = -\frac{1}{T_Y(y)}\frac{\partial^2 T_Y(y)}{\partial y^2} = \lambda^2 \tag{6.2a}$$

and solving the two resulting equations

$$\frac{\partial^2 T_X(x)}{\partial x^2} - \lambda^2\, T_X(x) = 0 \qquad\qquad (6.2b)$$

$$\frac{\partial^2 T_Y(y)}{\partial y^2} + \lambda^2\, T_Y(y) = 0 \qquad\qquad (6.2c)$$

subject to the problem's boundary conditions. λ is known as a separation constant. It can take on any of three separate values of λ^2, $-\lambda^2$ and $\lambda^2 = 0$. Each of these values of λ will result in a different solution of equations (6.2a).

For example, in the case of positive λ^2 in the region $0 < x < a$, equation (6.2b) has the solution

$$T_X(x) = A\cos(\lambda x) + B\sin(\lambda x) \qquad\qquad (6.3)$$

and in the region $0 < y < b$, equation (6.2c) has the solution

$$T_Y(y) = C\cosh(\lambda y) + D\sinh(\lambda y) \qquad\qquad (6.4)$$

Applying the boundary conditions

Because Laplace's equation is an elliptic, partial differential equation, a separation of variables solution requires it to have homogenous boundary conditions, the so called Dirichlet problem (Zill and Cullen, 1997). For example, if two parallel sides of the plate are insulated such that

1	$x = 0$	$\dfrac{\partial T}{\partial x}\Big	_{x=0} = 0$	$0 < y < b$
2	$x = a$	$\dfrac{\partial T}{\partial x}\Big	_{x=a} = 0$	$0 < y < b$
3	$y = 0$	$T(x, 0) = 0$	$0 < x < a$
4	$y = b$	$T(x, b) = f(x)$	$0 < x < a$

From equation (6.3) and boundary condition (1)

$$\frac{\partial T_X(x)}{\partial x}\bigg|_{x=0} = -\lambda A sin\,(\lambda \times 0) + \lambda B cos\,(\lambda \times 0) = 0$$

that is B = 0 and

$$T_X(x) = A cos\,(\lambda x) \qquad\qquad (6.3a)$$

From equation (3.3a) and boundary condition (2)

$$\frac{\partial T_X(x)}{\partial x}\bigg|_{x=a} = -\lambda A sin\,(\lambda a) = 0$$

which is true only when $\lambda = 0$ or when $\lambda a = n\pi$ that is when $\lambda = (n\pi/a)$. Note that when $\lambda = 0$ and n = 0, equation (6.2b) becomes

$$\frac{\partial^2 T_X(x)}{\partial x^2} = 0$$

which has the general solution,

$$T_X(x) = A + Bx \qquad\qquad (6.3b)$$

With boundary conditions (1) and (2) we can see that for $\lambda = 0$ and n = 0,

$$T_X(x) = A_o \qquad\qquad (6.3c)$$

and for $\lambda = (n\pi/a)$, n = 1, 2, 3, ...

$$T_X(x) = A_n \cos\left(\frac{n\pi}{a}\right)x \qquad\qquad (6.3d)$$

From equation (6.4) and boundary condition (3) and $\lambda > 0$

$$T_Y(0) = C cosh\,(\lambda \times 0) + D sinh\,(\lambda \times 0) = 0$$

that is C = 0 and

$$T_Y(y) = D sinh\,(\lambda y) \qquad\qquad (6.4a)$$

115

When $\lambda = 0$ and n = 0, equation (6.2c) becomes

$$\frac{\partial^2 T_Y(y)}{\partial y^2} = 0$$

which has the general solution,

$$T_Y(y) = C + Dy \qquad (6.4b)$$

With boundary conditions (3) we can see that for $\lambda = 0$, n = 0, and C = 0

$$T_Y(y) = D_o y \qquad (6.4c)$$

and for $\lambda = (n\pi/a)$, n = 1, 2, 3, ...

$$T_Y(y) = D_n \sinh\left(\frac{n\pi}{a}\right) y \qquad (6.4d)$$

From equations (6.1a), (6.3c), (6.3d), (6.4c) and (6.4d), the final solution obtained is

$$T(x,y) = T_X(x).T_Y(y) = A_o D_o y \qquad (6.5)$$

for $\lambda = 0$, n = 0, and

$$T(x,y) = T_X(x).T_Y(y) = \left[A_n \cos\left(\frac{n\pi}{a}x\right)\right].\left[D_n \sinh\left(\frac{n\pi}{a}y\right)\right] \quad (6.6)$$

for $\lambda = (n\pi/a)$, n = 1, 2, 3, ...

By the principle of superposition of solutions

$$T(x,y) = A_o D_o y + \sum_{n=1}^{\infty}\left[A_n \cos\left(\frac{n\pi}{a}x\right)\right].\left[D_n \sinh\left(\frac{n\pi}{a}y\right)\right] \quad (6.7)$$

We can simplify the constants and replace $A_o D_o$ with A_o and $A_n D_n$ with A_n so that

$$T(x, y) = A_o y + \sum_{n=1}^{\infty} A_n \cos\left(\frac{n\pi}{a} x\right) . \sinh\left(\frac{n\pi}{a} y\right) \qquad (6.8)$$

Using boundary condition (4) in equation (6.8), we get that

$$T(x, b) = f(x) = A_o b + \sum_{n=1}^{\infty} A_n \cos\left(\frac{n\pi}{a} x\right) . \sinh\left(\frac{n\pi}{a} b\right) \qquad (6.8a)$$

If we make

$$\frac{a_o}{2} = A_o b \quad and \quad a_n = A_n \sinh\left(\frac{n\pi}{a} b\right)$$

we find that equation (6.8a) is a half range expansion of f (x) as a cosine series from which we get that

$$2A_o b = \frac{2}{a} \int_0^a f(x) dx \quad or \quad A_o = \frac{1}{ab} \int_0^a f(x) dx \qquad (6.9)$$

and

$$A_n \sinh\left(\frac{n\pi}{a} b\right) = \frac{2}{a} \int_0^a f(x) \cos\left(\frac{n\pi}{a} x\right) dx$$

or

$$A_n = \frac{2}{a \sinh\left(\frac{n\pi}{a} b\right)} \int_0^a f(x) \cos\left(\frac{n\pi}{a} x\right) dx \qquad (6.9a)$$

Example 6.12: Since the separation of variables method can, only, be applied to situations with homogenous boundary conditions, is there a situation in which it can, still, be applied even though the boundary conditions are not homogenous?

Answer: The separation of variables method cannot, normally, be applied directly when the boundary conditions are not homogenous. In such situations, and where it is possible, the problem may be decomposed into two or more problems, each having homogenous boundaries. Each is then solved by separation of variables and the final solutions combined, as shown above, by the method of superposition of solutions.

For example, if instead of the boundary conditions we had above for the flat plate with parallel insulated boundaries we now have the following boundary conditions

1	$x = 0$	$T(0, y) = F(y)$	$0 < y < b$
2	$x = a$	$T(a, y) = G(y)$	$0 < y < b$
3	$y = 0$	$T(x, 0) = f(x)$	$0 < x < a$
4	$y = b$	$T(x, b) = g(x)$	$0 < x < a$

the Laplace's equation cannot be solved by separation of variables. But if it is stated as two problems

$$\frac{\partial^2 T_1}{\partial x^2} + \frac{\partial^2 T_1}{\partial y^2} = 0 \tag{6.10}$$

with boundary conditions

1	$x = 0$	$T_1(0, y) = 0$	$0 < y < b$
2	$x - a$	$T_1(a, y) = 0$	$0 < y < b$
3	$y = 0$	$T_1(x, 0) = f(x)$	$0 < x < a$
4	$y = b$	$T_1(x, b) = g(x)$	$0 < x < a$

and

$$\frac{\partial^2 T_2}{\partial x^2} + \frac{\partial^2 T_2}{\partial y^2} = 0 \tag{6.11}$$

with boundary conditions

1	$x = 0$	$T_2(0, y) = F(y)$	$0 < y < b$
2	$x = a$	$T_2(a, y) = G(y)$	$0 < y < b$
3	$y = 0$	$T_2(x, 0) = 0$	$0 < x < a$
4	$y = b$	$T_2(x, b) = 0$	$0 < x < a$

the solutions obtained by separation of variables can be shown to be (Zill and Cullen, 1997)

$$T_1(x, y) =$$

$$\sum_{n=1}^{\infty} \left[A_n \cosh\left(\frac{n\pi}{a}y\right) + B_n \sinh\left(\frac{n\pi}{a}y\right) \right] \sin\left(\frac{n\pi}{a}x\right) \quad (6.10a)$$

where

$$A_n = \frac{2}{a} \int_0^a f(x) \sin\left(\frac{n\pi}{a}x\right) dx \quad (6.10b)$$

$$B_n = \frac{1}{\sinh\left(\frac{n\pi}{a}b\right)} \left[\frac{2}{a} \int_0^a g(x) \sin\left(\frac{n\pi}{a}x\right) dx - A_n \cosh\left(\frac{n\pi}{a}b\right) \right] \quad (6.10c)$$

and

$$T_2(x, y) =$$

$$\sum_{n=1}^{\infty} \left[A_n \cosh\left(\frac{n\pi}{b}x\right) + B_n \sinh\left(\frac{n\pi}{b}x\right) \right] \sin\left(\frac{n\pi}{b}y\right) \quad (6.11a)$$

where

$$A_n = \frac{2}{b} \int_0^b F(y) \sin\left(\frac{n\pi}{b}y\right) dy \quad (6.11b)$$

$$B_n = \frac{1}{\sinh\left(\frac{n\pi}{b}a\right)} \left[\frac{2}{b} \int_0^b G(y) \sin\left(\frac{n\pi}{b}y\right) dy - A_n \cosh\left(\frac{n\pi}{b}a\right) \right] \quad (6.11c)$$

The final solution is, by superposition of solutions,

$$T(x, y) = T_1(x, y) + T_2(x, y) \quad (6.12)$$

Example 6.13: How would the separation of variables solution of the steady state, multi-dimensional Laplace's equation in heat conduction look like in cylindrical co-ordinates?

119

Answer: Interest in the two dimensional Laplace's heat conduction equation in polar co-ordinates arises in the case of cylinders and flat circular plates. For heat conduction in flat circular plates, interest is, usually, in the r and θ directions, while for cylinders, the interest is in the r and z directions.

Let us consider the case of steady state, two dimensional heat conduction in a flat, circular plate. The Laplace's equation, in the r and θ dimensions, can be stated as

$$\frac{\partial^2 T}{\partial r^2} + \frac{1}{r}\frac{\partial T}{\partial r} + \frac{1}{r^2}\frac{\partial^2 T}{\partial \theta^2} = 0 \qquad (6.13)$$

By separation of variables

$$T(r,\theta) = R(r).\Theta(\theta) \qquad (6.13a)$$

where R(r) is a function of r only and independent of θ while $\theta(\theta)$ is a function of θ only and independent of r. Equation (6.13a) can be differentiated to various orders as required

$$\frac{\partial T}{\partial r} = \Theta(\theta)\frac{\partial R(r)}{\partial r} \qquad (6.13b)$$

$$\frac{\partial^2 T}{\partial r^2} = \Theta(\theta)\frac{\partial^2 R(r)}{\partial r^2} \qquad (6.13c)$$

Similarly

$$\frac{\partial T}{\partial \theta} = R(r)\frac{\partial \Theta(\theta)}{\partial \theta} \qquad (6.13d)$$

$$\frac{\partial^2 T}{\partial \theta^2} = R(r)\frac{\partial^2 \Theta(\theta)}{\partial \theta^2} \qquad (6.13e)$$

When they are substituted into equation (6.13), we get

$$\Theta(\theta)\frac{\partial^2 R(r)}{\partial r^2} + \frac{\Theta(\theta)}{r}\frac{\partial R(r)}{\partial r} + \frac{R(r)}{r^2}\frac{\partial^2 \Theta(\theta)}{\partial \theta^2} = 0$$

That is

$$\frac{r^2}{R(r)}\frac{\partial^2 R(r)}{\partial r^2} + \frac{r}{R(r)}\frac{\partial R(r)}{\partial r} = -\frac{1}{\Theta(\theta)}\frac{\partial^2 \Theta(\theta)}{\partial \theta^2} \qquad (6.14)$$

Thus, the original equation has been transformed such that the left hand side and the right hand side of equation (6.14) are functions of one variable only. The solution is obtained by setting

$$\frac{r^2}{R(r)}\frac{\partial^2 R(r)}{\partial r^2} + \frac{r}{R(r)}\frac{\partial R(r)}{\partial r} = -\frac{1}{\Theta(\theta)}\frac{\partial^2 \Theta(\theta)}{\partial \theta^2} = \lambda^2 \qquad (6.14a)$$

and solving the two resulting equations, expressed as total, rather than partial, differentials, as

$$r^2\frac{d^2 R(r)}{dr^2} + r\frac{dR(r)}{dr} - \lambda^2 R(r) = 0 \qquad (6.14b)$$

$$\frac{d^2 \Theta(\theta)}{d\theta^2} + \lambda^2 \Theta(\theta) = 0 \qquad (6.14c)$$

Equation (6.14b), also known as the Cauchy-Euler equation (Zill & Cullen, 1997), has the solution

$$R(r) = \frac{A}{r^\lambda} + Br^\lambda \qquad (6.15)$$

while equation (6.14c) has the solution

$$\Theta(\theta) = C\cos(\lambda\theta) + D\sin(\lambda\theta) \qquad (6.16)$$

provided λ is an integer, possible only if $T(r, \theta+2\pi) = T(r, \theta)$.

It is usual to replace λ by n so that equation (6.15) becomes

$$R(r) = \frac{A}{r^n} + Br^n \qquad (6.15a)$$

while equation (6.16) becomes

$$\Theta(\theta) = C\cos(n\theta) + D\sin(n\theta) \tag{6.16a}$$

The final solution is then

$$T(r,\theta) = R(r).\Theta(\theta) = \left(\frac{A}{r^n} + Br^n\right)(C\cos n\theta + D\sin n\theta) \tag{6.17}$$

Applying the boundary conditions

If a is the radius of the circular plate, then for $r \le a$ (Dirichlet's interior problem), A = 0 if T is not to be infinite when $r = 0$. Then

$$T(r,\theta) = r^n(A_n \cos n\theta + B_n \sin n\theta) \tag{6.17a}$$

For $r \ge a$ (Dirichlet's exterior problem), B = 0 if T is not to be infinite when $r \to \infty$. Then

$$T(r,\theta) = \frac{1}{r^n}(C_n \cos n\theta + D_n \sin n\theta) \tag{6.17b}$$

Equations (6.17a) and (6.17b) are each, a solution to the problem. If, for example, we use the boundary condition that at $r = a$

$$T(r,\theta) = T(a,\theta) = f(\theta)$$

then for $r \le a$

$$f(\theta) = \sum_{n=0}^{\infty} a^n(A_n \cos n\theta + B_n \sin n\theta) \tag{6.17c}$$

For $r \ge a$

$$f(\theta) = \sum_{n=0}^{\infty} \frac{1}{a^n}(C_n \cos n\theta + D_n \sin n\theta) \tag{6.17d}$$

for n = λ > 0. When n = λ = 0 and r = 0, by similar analysis as before

$$r^2 \frac{d^2 R(r)}{dr^2} + r\frac{dR(r)}{dr} = 0 \tag{6.18}$$

$$\frac{d^2\Theta(\theta)}{d\theta^2} = 0 \qquad\qquad (6.19)$$

with solutions

$$R(r) = E + F\ln{(r)} \qquad\qquad (6.18a)$$

and

$$\Theta(\theta) = G\theta + H \qquad\qquad (6.19a)$$

where E, F, G and H are constants.. Hence for R(r) not to be infinite at r = 0, F has to be zero. Because T(0, θ +2π) = T(0,θ), G has, also, got to be zero. Thus

$$T(0,\theta) = R(0).\Theta(\theta) = E.H = A_o \qquad\qquad (6.20)$$

It can be shown in Fourier series analysis that for $r \le a$

$$T(r,\theta) = \frac{\alpha_o}{2} + \sum_{n=1}^{\infty} \left(\frac{r}{a}\right)^n [\alpha_n \cos{(n\theta)} + \beta_n \sin{(n\theta)}] \qquad (6.21)$$

where

$$\alpha_o = \frac{1}{\pi}\int_{-\pi}^{\pi} f(\lambda)d\lambda \qquad\qquad (6.21a)$$

$$\alpha_n = \frac{1}{\pi}\int_{-\pi}^{\pi} f(\lambda)\cos(n\lambda)\,d\lambda \qquad\qquad (6.21b)$$

$$\beta_n = \frac{1}{\pi}\int_{-\pi}^{\pi} f(\lambda)\,\text{ain}(n\lambda)\,d\lambda \qquad\qquad (6.21c)$$

$$A_o = \frac{\alpha_o}{2} \qquad\qquad (6.21d)$$

$$A_n = \frac{\alpha_n}{a^n} \qquad\qquad (6.21e)$$

$$B_n = \frac{\beta_n}{a^n} \qquad\qquad (6.21f)$$

and n = 1, 2, 3, …..

Similarly, for $r \geq a$

$$T(r,\theta) = \frac{a_o}{2} + \sum_{n=1}^{\infty} \left(\frac{a}{r}\right)^n [a_n \cos(n\theta) + \beta_n \sin(n\theta)] \qquad (6.22)$$

You can imagine that there would be several types of solutions for different heat conduction configurations such as the semi-circular plate, the concentric plate, triangular plate, etc. The solution obtained would depend on the boundary conditions.

Example 6.14: How would the separation of variables solution of the steady state, multi-dimensional Laplace's equation in heat conduction, in a cylinder, look like in cylindrical co-ordinates?

Answer: This is the case in which the variation of temperature, T, is along the r and z directions. This situation is common in heat conduction in cylindrical objects. The Laplace's equation becomes, in the r and z directions,

$$\frac{\partial^2 T}{\partial r^2} + \frac{1}{r}\frac{\partial T}{\partial r} + \frac{\partial^2 T}{\partial z^2} = 0 \qquad (6.23)$$

Separation of variables method requires that

$$T(r,z) = R(r).Z(z)$$

That is

$$Z(z)\frac{d^2 R(r)}{dr^2} + \frac{Z(z)}{r}\frac{dR(r)}{dr} + R(r)\frac{d^2 Z(z)}{dz^2} = 0 \qquad (6.24)$$

Equation (6.24) can be solved using a positive separation of variables constant with results similar to that obtained for temperature variation in the r and θ directions. Because we do not expect the solution to be periodic in z, however, a negative value of the separation constant can be used to get another type of solution. Whichever one is used will depend on how applicable it is to the system under consideration.

For example, if we use a negative separation constant, we find that

$$\frac{1}{R(r)}\frac{d^2R(r)}{dr^2} + \frac{1}{rR(r)}\frac{dR(r)}{dr} = -\frac{1}{Z(z)}\frac{d^2Z(z)}{dz^2} = -\lambda^2 \qquad (6.25)$$

from which

$$r\frac{d^2R(r)}{dr^2} + \frac{dR(r)}{dr} + \lambda^2 rR(r) = 0 \qquad (6.25a)$$

$$\frac{d^2Z(z)}{dz^2} - \lambda^2 Z(z) = 0 \qquad (6.25b)$$

Equation (6.25a) has the general solution

$$R(r) = AJ_0(\lambda r) + BY_0(\lambda r) \qquad (6.25c)$$

while equation (6.25b) has the solution

$$Z(z) = C\cosh(\lambda z) + D\sinh(\lambda z) \qquad (6.25d)$$

The general solution is

$$T(r,z) = R(r).Z(z)$$

$$= [AJ_0(\lambda r) + BY_0(\lambda r)][C\cosh(\lambda z) + D\sinh(\lambda z)] \qquad (6.26)$$

$J_0(\lambda r)$ and $Y_0(\lambda r)$ are Bessel's functions of the first and second kinds, respectively, of order zero. For any order, v,

$$J_v(\lambda r) = \sum_{n=0}^{\infty} \frac{(-1)^n}{n!\,\Gamma(1+v+n)}\left(\frac{\lambda r}{2}\right)^{2n+v} \qquad (6.27)$$

$$Y_v(\lambda r) = \frac{\cos(v\pi)J_v(\lambda r) - J_{-v}(\lambda r)}{\sin(v\pi)} \qquad (6.28)$$

If the boundary conditions are, for example,

$z = 0$	$T(r, 0) = 0$	$0 < r < a$
$z = b$	$T(r, b) = T_0$	$0 < r < a$
$r = a$	$T(a, z) = 0$	$0 < z < b$

$$(6.29)$$

then, at $r = 0$, from equation (6.25c), $B = 0$. That is

$$R(r) = AJ_o(\lambda_n r) \tag{6.30}$$

At $r = a$, $T(a, z) = 0$, $R(a) = 0$ or from equation (6.25c)

$$AJ_o(\lambda_n r) = 0 \tag{6.31}$$

which defines the values of λ_n.

Since $T(r,0) = 0$ at $z = 0$, $C = 0$ from equation (6.25d). That is

$$Z(z) = Dsinh\,(\lambda_n z) \tag{6.32}$$

Since $T(r, z) = R(r).Z(z)$ we obtain that

$$T_n(r, z) = A_n J_o(\lambda_n r) D_n sinh\,(\lambda_n z) \tag{6.33}$$

We can replace $A_n D_n$ by A_n to get the general solution

$$T(r, z) = \sum_{n=1}^{\infty} A_n \sinh(\lambda_n z) J_o(\lambda_n r) \tag{6.34}$$

From $T(r, b) = T_0$ at $z = b$

$$T_o = \sum_{n=1}^{\infty} A_n \sinh(\lambda_n b) J_o(\lambda_n r) \tag{6.34a}$$

It is found from the Fourier-Bessel series that

$$A_n = \frac{T_o}{\lambda_n \sinh(b\lambda_n) J_1(a\lambda_n)} \tag{6.35}$$

so that the final solution becomes

$$T(r,z) = T_o \sum_{n=1}^{\infty} \frac{\sinh(\lambda_n z) J_0(\lambda_n r)}{\lambda_n \sinh(b\lambda_n) J_1(a\lambda_n)} \qquad (6.36)$$

Example 6.15: How would the separation of variables solution of the steady state, multi-dimensional Laplace's equation in heat conduction look like in spherical co-ordinates?

Answer: The statement of the Laplace's equation, in spherical co-ordinates, is

$$\frac{1}{r^2} \frac{\partial}{\partial r}\left(r^2 \frac{\partial T}{\partial r}\right) + \frac{1}{r^2 \sin\theta} \frac{\partial}{\partial\theta}\left(\sin\theta \frac{\partial T}{\partial\theta}\right) + \frac{1}{r^2 \sin^2\theta} \frac{\partial^2 T}{\partial\varphi^2} = 0$$

Analytical solution of this equation is complex for all but the simplest cases. If, for example, the temperature does not vary in the φ direction, the separation of variables form of the equation is

$$\frac{r^2}{R(r)} \frac{d^2 R(r)}{dr^2} + \frac{2r}{R(r)} \frac{dR(r)}{dr} = -\frac{1}{\Theta(\theta)} \frac{d^2\Theta(\theta)}{d\theta^2} - \frac{\cot\theta}{\Theta(\theta)} \frac{d\Theta(\theta)}{d\theta})$$
$$= -\lambda^2 \qquad (6.37$$

The equations to be solved are

$$r^2 \frac{d^2 R(r)}{dr^2} + 2r \frac{dR(r)}{dr} - \lambda^2 R(r) = 0 \qquad (6.38)$$

and

$$\sin\theta \frac{d^2\Theta(\theta)}{d\theta^2} + \cos\theta \frac{d\Theta(\theta)}{d\theta} + \lambda^2\Theta(\theta)\sin\theta = 0 \qquad (6.39)$$

If the boundary conditions are such that $T(a,0) = f(\theta)$ for $0 < r < a$ and $0 < \theta < \pi$, the solution is (Zill & Cullen, 1997)

$$T(r,\theta) = \sum_{n=0}^{\infty} \left[\frac{2n+1}{2} \int_{0}^{\pi} f(\theta)\, P_n(\cos\theta)\sin\theta\, d\theta \right] \left(\frac{r}{a}\right)^n P_n(\cos\theta) \quad (6.40)$$

where $P_n(\cos\theta)$ is the Legendre polynomial.

Example 6.16: How would the analytical solutions of the steady state, multi-dimensional Poisson's equation in heat conduction look like?

Answer: While analytical solutions are the first method of choice for one dimensional and simple two dimensional problems, they can get complicated so that they are not often the methods of choice unless other methods are less attractive. The analytical solution of the Poisson equation is one such example. Whether expressed in Cartesian

$$\frac{Q_V}{k} + \frac{\partial^2 T}{\partial x^2} + \frac{\partial^2 T}{\partial y^2} + \frac{\partial^2 T}{\partial z^2} = 0$$

cylindrical polar

$$\frac{Q_V}{k} + \frac{\partial^2 T}{\partial r^2} + \frac{1}{r}\frac{\partial T}{\partial r} + \frac{1}{r^2}\frac{\partial^2 T}{\partial \theta^2} + \frac{\partial^2 T}{\partial z^2} = 0$$

or spherical co-ordinates

$$\frac{Q_V}{k} + \frac{1}{r^2}\frac{\partial}{\partial r}\left(r^2\frac{\partial T}{\partial r}\right) + \frac{1}{r^2\sin\theta}\frac{\partial}{\partial\theta}\left(\sin\theta\frac{\partial T}{\partial\theta}\right) + \frac{1}{r^2\sin^2\theta}\frac{\partial^2 T}{\partial\varphi^2} = 0$$

the Poisson equation is not homogenous and, thus, is not easily solved analytically by simple separation of variables. A common mathematical procedure, of changing the dependent variable to be the sum of two dependent variables one of which is a function of one variable and the other of two variables in such a way that the new variable satisfies both a homogenous partial differential equation and a homogenous boundary condition (Zill & Cullen, 1997) can be used, however. This is beyond the scope of this book.

Example 6.17: What of three dimensional solutions of the steady state, multi-dimensional heat conduction?

Answer: When three dimensions are considered simultaneously such as the x, y, z in Cartesian co-ordinates, the r, θ, z in cylindrical co-ordinates or the r, θ, φ in spherical co-ordinates, the separated variables, for example, in cylindrical co-ordinates are

$$T(r,\theta,z) = R(r).\Theta(\theta).Z(z) \tag{6.41}$$

The Laplace's equation takes the form

$$\frac{1}{R(r)}\frac{d^2R(r)}{dr^2} + \frac{1}{rR(r)}\frac{dR(r)}{dr} + \frac{1}{r^2\Theta(\theta)}\frac{d^2\Theta(\theta)}{d\theta^2} + \frac{1}{Z(z)}\frac{d^2Z(z)}{dz^2}$$
$$= 0 \qquad (6.42)$$

By letting

$$\frac{1}{\Theta(\theta)}\frac{d^2\Theta(\theta)}{d\theta^2} = -\beta \tag{6.43}$$

and

$$\frac{1}{Z(z)}\frac{d^2Z(z)}{dz^2} = \alpha \tag{6.44}$$

the equation to be solved is

$$r^2\frac{d^2R(r)}{dr^2} + r\frac{dR(r)}{dr} + (\alpha r^2 - \beta)R(r) = 0 \tag{6.45}$$

the Bessel's equation. The solution of this equation yields the so called Bessel's function whose treatment here is beyond the scope of this book.

Example 6.18: Outline the finite difference formulations of steady state heat conduction in two dimensions.

Answer: The finite difference formulations, in Cartesian co-ordinates, of steady state heat conduction in two dimensions are shown in Table 6.3

Table 6.3: Finite Difference Formulation of Steady State Heat Conduction in Two Dimensions (Welty, 1978)

Differential Equation	Finite Difference Formulation
Laplace Equation $$\frac{\partial^2 T}{\partial x^2} + \frac{\partial^2 T}{\partial y^2} = 0$$	General Form $$\frac{T_{i-1,j} - 2T_{i,j} + T_{i+1,j}}{(\Delta x)^2}$$ $$+ \frac{T_{i,j-1} - 2T_{i,j} + T_{i,j+1}}{(\Delta y)^2} = 0$$ Explicit Form: if $\Delta x = \Delta y$ $$T_{i,j} = \frac{T_{i-1,j} + T_{i+1,j} + T_{i,j-1} + T_{i,j+1}}{4}$$
Poisson Equation $$\frac{\partial^2 T}{\partial x^2} + \frac{\partial^2 T}{\partial y^2} + \frac{q}{k} = 0$$	General Form $$\frac{T_{i-1,j} - 2T_{i,j} + T_{i+1,j}}{(\Delta x)^2}$$ $$+ \frac{T_{i,j-1} - 2T_{i,j} + T_{i,j+1}}{(\Delta y)^2} + \frac{q}{k}$$ $$= 0$$ Explicit Form; if $\Delta x = \Delta y$

$$T_{i,j} = \frac{T_{i-1,j} + T_{i+1,j} + T_{i,j-1} + T_{i,j+1}}{4}$$
$$+ \frac{q(\Delta x)^2}{4k}$$

Example 6.19: How would these finite difference formulations be applied in practice?

Answer: Consider a numerical solution of the two dimensional Laplace equation, in Cartesian co-ordinates. We can replace the partial differential equation by a difference equation using the finite difference formulations we derived. Thus, since $\Delta x = h$, if we make $\Delta x = \Delta y$,

$$\frac{\partial^2 T}{\partial x^2} + \frac{\partial^2 T}{\partial y^2} = \frac{T_{i-1,j} - 2T_{i,j} + T_{i+1,j}}{(h)^2} + \frac{T_{i,j-1} - 2T_{i,j} + T_{i,j+1}}{(h)^2} = 0$$

This is the equivalent of

$$T(x + h) + T(x, y + h) + T(x - h, y) + T(x, y - h) - 4T(x, y)$$
$$= 0$$

The notation of ij is more convenient to us. Thus

$$T_{i+1,j} + T_{i,j+1} + T_{i-1,j} + T_{i,j-1} - 4T_{i,j} = 0 \qquad (6.46)$$

is the equation to be solved implicitly. The explicit form is, from equation (6.46)

$$T_{i,j} = \frac{T_{i+1,j} + T_{i,j+1} + T_{i-1,j} + T_{i,j-1}}{4} \qquad (6.47)$$

To get a better feel of what we are talking about, it is useful to draw the grid of the points described by these notations.

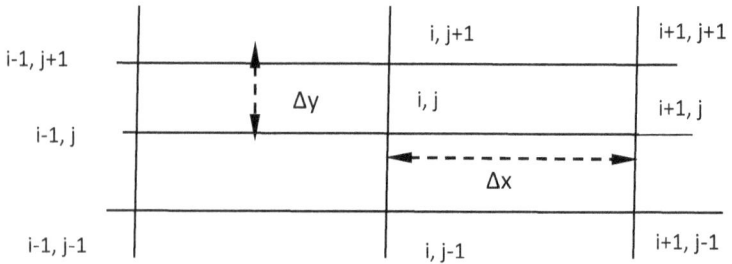

In a real problem, the points, also called nodes, mesh or lattice points, can be visualised on an *x-y* graph on which the boundaries of the surface, under consideration, can be located.

Thus, the rectangular plate shown below has the corner points, clockwise, from the bottom left corner of the plate, T(h, h), T(h, 4h), T(4h, 4h) and T(4h,h) or T_{11}, T_{14}, T_{44} and T_{41}. If, for example, T_{11} is equivalent to x = 3.5, y = 3.5 and T_{41} to x = 5, y = 5. Then h = (5-3.5)/3 = 0.5.

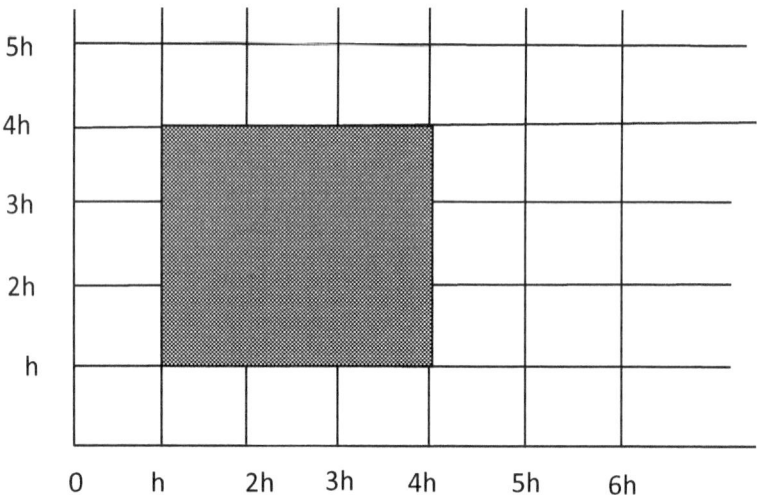

Suppose, also, that the boundary conditions are such that T_{11}, T_{12}, T_{13}, T_{14} and T_{41}, T_{42}, T_{43}, T_{44}, are known and that what needs to be determined are the interior points, T_{22}, T_{23}, T_{32}, T_{33}. Usually, the number of interior points is $(n-1)^2$. Since equation (6.46) applies to each node or lattice point, we can use it to evaluate T_{ij} as follows

$$T_{i+1,j} + T_{i,j+1} + T_{i-1,j} + T_{i,j-1} - 4T_{i,j} = 0 \qquad (6.46)$$

At node, T_{22}

$$T_{32} + T_{23} + T_{12} + T_{21} - 4T_{22} = 0 \qquad (6.46a)$$

At node, T_{23}

$$T_{33} + T_{24} + T_{13} + T_{22} - 4T_{23} = 0 \qquad (6.46b)$$

At node, T_{32}

$$T_{42} + T_{33} + T_{22} + T_{31} - 4T_{32} = 0 \qquad (6.46c)$$

At node, T_{33}

$$T_{43} + T_{34} + T_{23} + T_{32} - 4T_{33} = 0 \qquad (6.46d)$$

Rearranging the equations

$$-4T_{22} + T_{23} + T_{32} = -T_{12} - T_{21}$$
$$T_{22} - 4T_{23} + T_{33} = -T_{13} - T_{24}$$
$$T_{22} - 4T_{32} + T_{33} == -T_{31} - T_{42}$$
$$T_{23} + T_{32} - 4T_{33} = -T_{34} - T_{43}$$

This can be expressed as a matrix

$$
\begin{pmatrix}
-4 & 1 & 1 & 0 \\
1 & -4 & 0 & 1 \\
1 & 0 & -4 & 1 \\
0 & 1 & 1 & -4
\end{pmatrix}
\begin{pmatrix}
T_{22} \\
T_{23} \\
T_{32} \\
T_{33}
\end{pmatrix}
=
\begin{pmatrix}
-T_{12} - T_{21} \\
-T_{13} - T_{24} \\
-T_{31} - T_{42} \\
-T_{34} - T_{43}
\end{pmatrix}
\qquad (6.47)
$$

which can be solved because the right hand side is known from the boundary conditions.

Note that if n is increased from 3 to 4 as shown in the figure below, the number of interior points to be solved for becomes $(4-1)^2 = 9$ requiring 9 equations in order to solve for the 9 unknowns.

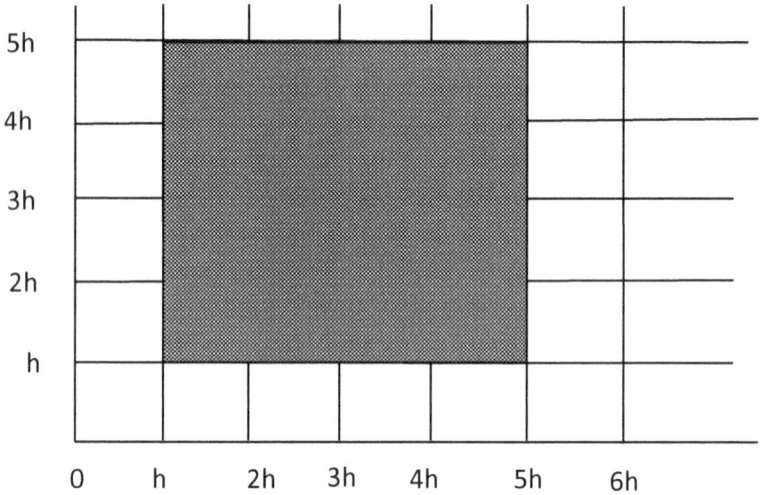

Example 6.20: How would these finite difference formulations look like when applied to the Poisson's equation?

Answer: The analysis is similar to that for the Laplace equation. We can replace the partial difference equation by a difference equation using the finite difference formulations we derived. Thus, since $\Delta x = h$, if we make $\Delta x = \Delta y$,

$$\frac{\partial^2 T}{\partial x^2} + \frac{\partial^2 T}{\partial y^2} + \frac{q}{k} = \frac{T_{i-1,j} - 2T_{i,j} + T_{i+1,j}}{h^2} + \frac{T_{i,j-1} - 2T_{i,j} + T_{i,j+1}}{h^2} + \frac{q}{k} = 0$$

This is the equivalent of

$$\frac{T_{i-1,j} - 2T_{i,j} + T_{i+1,j}}{h^2} + \frac{T_{i,j-1} - 2T_{i,j} + T_{i,j+1}}{h^2} + \frac{q}{k} = 0 \qquad (6.48)$$

in general form, and

$$T_{i,j} = \frac{T_{i-1,j} + T_{i+1,j} + T_{i,j-1} + T_{i,j+1}}{4} + \frac{qh^2}{4k} \qquad (6.49)$$

in explicit form.

Example 6.21: Show that

$$\theta(x, y) = 2\theta_0 \sum_{n=1}^{\infty} \left(\frac{\sin \lambda_n L}{\lambda_n L + \sin(\lambda_n L) \cos(\lambda_n L)} \right) e^{-\lambda_n x} \cos(\lambda_n y)$$

is a solution of the Laplace's equation for heat transfer

$$\frac{\partial^2 \theta(x, y)}{\partial x^2} + \frac{\partial^2 \theta(x, y)}{\partial y^2} = 0$$

Answer: Since

$$\frac{\sin \lambda_n L}{\lambda_n L + \sin(\lambda_n L) \cos(\lambda_n L)}$$

is constant with respect to x and y, let it be represented by A. The given equation then becomes

$$\theta(x, y) = 2\theta_0 \sum_{n=1}^{\infty} A e^{-\lambda_n x} \cos(\lambda_n y) \qquad (1)$$

Differentiating equation (1) with respect to x

$$\frac{\partial \theta(x, y)}{\partial x} = 2\theta_0 \sum_{n=1}^{\infty} -A\lambda_n e^{-\lambda_n x} \cos(\lambda_n y) \qquad (2)$$

and again

$$\frac{\partial^2 \theta(x, y)}{\partial x^2} = 2\theta_0 \sum_{n=1}^{\infty} A(\lambda_n)^2 e^{-\lambda_n x} \cos(\lambda_n y) \qquad (3)$$

Similarly, differentiating equation (1) with respect to y

$$\frac{\partial \theta(x, y)}{\partial y} = 2\theta_0 \sum_{n=1}^{\infty} -A\lambda_n e^{-\lambda_n x} \sin(\lambda_n y) \qquad (4)$$

and again

$$\frac{\partial^2\theta(x,y)}{\partial y^2} = -2\theta_o \sum_{n=1}^{\infty} A(\lambda_n)^2\, e^{-\lambda_n x}\cos(\lambda_n y) \qquad (5)$$

Adding equations (3) and (5)

$$\frac{\partial^2\theta(x,y)}{\partial x^2} + \frac{\partial^2\theta(x,y)}{\partial y^2} =$$

$$2\theta_o \sum_{n=1}^{\infty} A(\lambda_n)^2\, e^{-\lambda_n x}\cos(\lambda_n y) - 2\theta_o \sum_{n=1}^{\infty} A(\lambda_n)^2\, e^{-\lambda_n x}\cos(\lambda_n y)$$

$$= 0 \quad Ans$$

Example 6.22: The analytical solution by, the method of separation of variables, of the two dimensional steady heat conduction equation.

$$\frac{\partial^2 T}{\partial x^2} + \frac{\partial^2 T}{\partial y^2} = 0$$

is given by

$$T - T_\infty = \frac{4(T_o - T_\infty)}{\pi} \sum_{n=1,3,}^{\infty} \left[\frac{1}{n}(-1)^{\frac{n-1}{2}} e^{-\lambda_n x}\cos(\lambda_n y)\right]$$

with boundary conditions

$$T(0,y) = F(y); \quad T(x,\pm1) = T_\infty; \quad T(\infty,y) = T_\infty$$

where $\lambda_n = \frac{n\pi}{2L}$ and L is the half thickness of the fin.
Determine the temperature distribution in a rectangular fin of uniform thickness 10mm, of total length 10mm, given that $T_\infty = 100\,C$ and $T_0 = 300$ C.

Answer: A section of the fin may be visualised as shown below

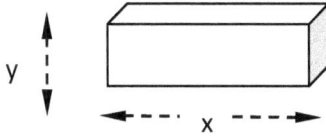

From the data given

$$\frac{4(T_o - T_\infty)}{\pi} = \frac{4(300 - 100)}{\pi} = 254.55$$

$$\lambda_n = \frac{n\pi}{2L} = 314.29n$$

The given equation can now be simplified to

$$T = 100 + 254.55 \sum_{n=1,3,}^{\infty} \left[\frac{1}{n}(-1)^{\frac{n-1}{2}} e^{-314.29nx} \cos(314.29ny)\right] (1)$$

What is required is a number of calculations of temperature, T, at enough points in the fin to give a reasonable picture of the temperature distribution. We can, also, see from the equation that there is no temperature variation in the z-direction.

Let us choose three values of x and five values of y co-ordinates to be x = 0, 0.005, 0.010 at values of y = 0.001, 0.002, 0.003, 0.004, 0.005. In an exaggerated diagram of the x-y face of the fin, the points we have chosen may be located in that plane of the fin as shown in the diagram below.

Remembering that n must have only odd numbered values, we can evaluate equation (1) in a spreadsheet in a computer. In order to illustrate how the calculations are done, however, we shall evaluate the groups in equation (1) manually and separately and compute the temperature for one or two points in the grid, as illustration, before doing the entire calculation on a spreadsheet in a computer.

137

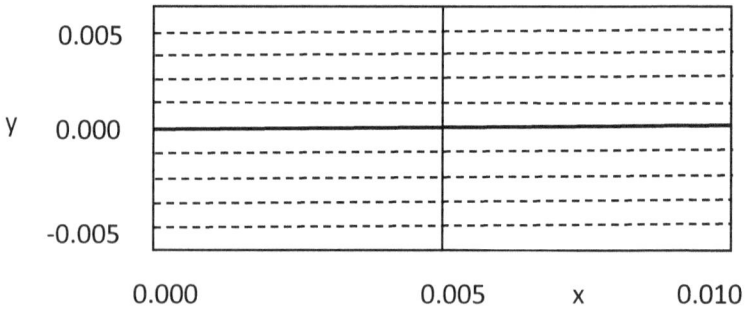

Thus the computed values of $e^{-314.29nx}$ and $\cos(314.29ny)$ are tabulated below for various values of x, y and n

Table of Values of $e^{-314.29nx}$

n/x	0	0.005	0.010
1	1.0000	0.2077	0.0432
3	1.0000	0.0090	0.0001
5	1.0000	0.0004	0.0000
7	1.0000	0.0000	0.0000
9	1.0000	0.0000	0.0000
11	1.0000	0.0000	0.0000
13	1.0000	0.0000	0.0000
15	1.0000	0.0000	0.0000
17	1.0000	0.0000	0.0000
19	1.0000	0.0000	0.0000

Table of Values of $\cos(314.29ny)$

n/y	0	0.001	0.002	0.003	0.004	0.005
1	1.0000	0.9510	0.8089	0.5875	0.3085	-0.0007
3	1.0000	0.5875	-0.3098	-0.9514	-0.8081	0.0020
5	1.0000	-0.0007	-1.0000	0.0020	1.0000	-0.0033
7	1.0000	-0.5885	-0.3073	0.9502	-0.8112	0.0046
9	1.0000	-0.9514	0.8104	-0.5906	0.3135	-0.0059

138

11	1.0000	-0.9506	0.8073	-0.5843	0.3035	0.0072
13	1.0000	-0.5864	-0.3122	0.9526	-0.8050	-0.0085
15	1.0000	0.0020	-1.0000	-0.0059	1.0000	0.0098
17	1.0000	0.5896	-0.3048	-0.9490	-0.8142	-0.0111
19	1.0000	0.9518	0.8119	0.5938	0.3185	0.0124

$$\text{Values of } \frac{1}{n}(-1)^{\frac{n-1}{2}} e^{-314.29nx} \cos(314.29ny) \text{ at } x = 0\text{m}$$

n/y	0	0.001	0.002	0.003	0.004	0.005
1	1.0000	0.9510	0.8089	0.5875	0.3085	-0.0007
3	-0.3333	-0.1958	0.1033	0.3171	0.2694	-0.0007
5	0.2000	-0.0001	-0.2000	0.0004	0.2000	-0.0007
7	-0.1429	0.0841	0.0439	-0.1357	0.1159	-0.0007
9	0.1111	-0.1057	0.0900	-0.0656	0.0348	-0.0007
11	-0.0909	0.0864	-0.0734	0.0531	-0.0276	-0.0007
13	0.0769	-0.0451	-0.0240	0.0733	-0.0619	-0.0007
15	-0.0667	-0.0001	0.0667	0.0004	-0.0667	-0.0007
17	0.0588	0.0347	-0.0179	-0.0558	-0.0479	-0.0007
19	-0.0526	-0.0501	-0.0427	-0.0313	-0.0168	-0.0007
SUM	0.7605	0.7592	0.7547	0.7433	0.7077	-0.0066

$$\text{Values of } \frac{1}{n}(-1)^{\frac{n-1}{2}} e^{-314.29nx} \cos(314.29ny) \text{ at } x = 0.005m$$

n/y	0	0.001	0.002	0.003	0.004	0.005
1	0.1039	0.0988	0.0840	0.0610	0.0320	-0.0001
3	0.0015	0.0009	-0.0005	-0.0014	-0.0012	0.0000
5	0.0000	0.0000	0.0000	0.0000	0.0000	0.0000
7	0.0000	0.0000	0.0000	0.0000	0.0000	0.0000
9	0.0000	0.0000	0.0000	0.0000	0.0000	0.0000
11	0.0000	0.0000	0.0000	0.0000	0.0000	0.0000

13	0.0000	0.0000	0.0000	0.0000	0.0000	0.0000
15	0.0000	0.0000	0.0000	0.0000	0.0000	0.0000
17	0.0000	0.0000	0.0000	0.0000	0.0000	0.0000
19	0.0000	0.0000	0.0000	0.0000	0.0000	0.0000
SUM	0.1054	0.0997	0.0835	0.0596	0.0309	-0.0001

$$\text{Values of } \frac{1}{n}(-1)^{\frac{n-1}{2}} e^{-314.29nx} \cos(314.29ny) \text{ at } x = 0.010m$$

n/y	0	0.001	0.002	0.003	0.004	0.005
1	0.0216	0.0205	0.0175	0.0127	0.0067	0.0000
3	0.0000	0.0000	0.0000	0.0000	0.0000	0.0000
5	0.0000	0.0000	0.0000	0.0000	0.0000	0.0000
7	0.0000	0.0000	0.0000	0.0000	0.0000	0.0000
9	0.0000	0.0000	0.0000	0.0000	0.0000	0.0000
11	0.0000	0.0000	0.0000	0.0000	0.0000	0.0000
13	0.0000	0.0000	0.0000	0.0000	0.0000	0.0000
15	0.0000	0.0000	0.0000	0.0000	0.0000	0.0000
17	0.0000	0.0000	0.0000	0.0000	0.0000	0.0000
19	0.0000	0.0000	0.0000	0.0000	0.0000	0.0000
SUM	0.0216	0.0205	0.0175	0.0127	0.0066	0.0000

A value of T is calculated from equation (1) and from the values in the tables from

$$T = 100 + 254.55 \sum_{n=1,3,}^{\infty} \left[\frac{1}{n}(-1)^{\frac{n-1}{2}} e^{-314.29nx} \cos(314.29ny) \right] (1)$$

Thus, at $x = 0$, $y = 0$ and odd $n = 1$ to 19,

$$T = 100 + 254.55 \sum_{n=1,3,}^{\infty} \left[\frac{1}{n}(-1)^{\frac{n-1}{2}} x\, 1\, x\, 1 \right]$$

$$= 100 + 254.55 \left[1 - \frac{1}{3} + \frac{1}{5} - \frac{1}{7} + \frac{1}{9} - \frac{1}{11} + \frac{1}{13} - \frac{1}{15} + \frac{1}{17} - \frac{1}{19}\right]$$

$$= 100 + 254.55 \times 0.7605 = 293.59 \ C$$

Similarly, at $x = 0$, $y = 0.001$ and from equation (1)

$$T = 100 + 254.55 \sum_{n=1,3,}^{\infty} \left[\frac{1}{n}(-1)^{\frac{n-1}{2}} x \ 1 \ x \ \cos(314.29ny)\right]$$

That is

$$T = 100 + 254.55 \left[1 \ x \ 1 \ x \ 0.9510 - \frac{0.5875}{3} - \frac{0.0007}{5} + \frac{0.5885}{7}\right.$$
$$\left. - \frac{0.9514}{9} + \frac{0.9506}{11} - \frac{0.5864}{13} - \frac{0.0020}{15} + \frac{0.5896}{17}\right.$$
$$\left. - \frac{0.9518}{19}\right]$$

$$= 100 + 254.55 \times 0.7592 = 293.25 \ C$$

Note that, if we do not sum n to infinity or to a sufficiently large number, our estimates of T will be in error. For example, summing to n = 17 instead of to, at least $n = 19$, gives estimates of T of 307 C at $x = 0$, $y = 0$ and of 306.01 C at $x = 0$, $y = 0.001$ both of which exceed the given mother surface temperature of 300 C.

Thus T at various x and y are calculated and tabulated below.

Values of $\sum_{n=1,3,}^{\infty} \left[\frac{1}{n}(-1)^{\frac{n-1}{2}} e^{-314.29nx} \cos(314.29ny)\right]$

x/y	0	0.001	0.002	0.003	0.004	0.005
0	0.7605	0.7592	0.7547	0.7433	0.7077	-0.0066
0.005	0.1054	0.0997	0.0835	0.0596	0.0309	-0.0001
0.010	0.0216	0.0205	0.0175	0.0127	0.0066	0

Values of Computed Temperature Distribution on the x-y Face of Fin

x/y	0	0.001	0.002	0.003	0.004	0.005
0	293.59	293.25	292.11	289.21	280.15	98.32
0.005	126.83	125.38	121.25	115.17	107.87	99.97
0.010	105.50	105.22	104.45	103.23	101.68	100.00

Fig. 6.1: Temperature Profile on x-y Face of Fin

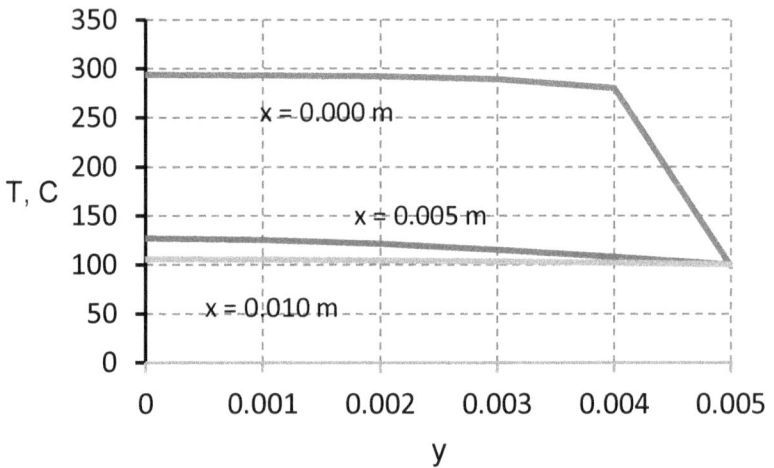

Example 6.23: The finite difference formulation of the two dimensional steady state heat conduction equation

$$\frac{\partial^2 T}{\partial x^2} + \frac{\partial^2 T}{\partial y^2} = 0$$

is given by, when $\Delta x = \Delta y$

$$T_{i,j} = \frac{T_{i-1,j} + T_{i+1,j} + T_{i,j-1} + T_{i,j+1}}{4}$$

where *ij* is a nodal point representing x and y co-ordinates while the heat flux between any two planes 1 and 2 is given by

$$q = k\frac{\Delta y}{\Delta x}\sum_{j=1}^{M}(T_{1j} - T_{2j})$$

where M = number of nodal elements.

Determine the steady state temperature distribution and total heat flux for a two dimensional steel plate whose one edge is maintained at 750 K while its three other edges are maintained at 330K. Take the thermal conductivity of the steel to be 39 W/m.K.

Answer: The plate and its division into nodal elements may be shown schematically as below

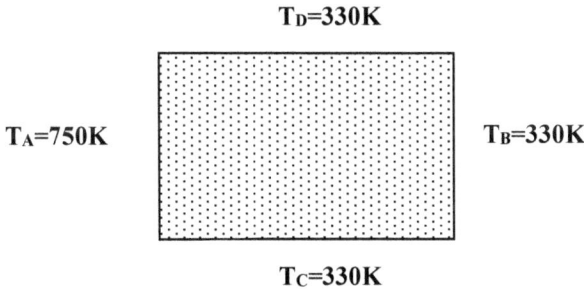

$T_D=330K$

$T_A=750K$ $T_B=330K$

$T_C=330K$

To have a manageable hand calculation example, we have divided the plate into three sections, each consisting of three elements in the x and y directions, as shown below. This means that the number of interior points will be $2^{(3-1)/2} = 4$. These points are T_{11}, T_{12}, T_{22} and T_{21}.

It is usual to solve the difference formulation by means of a computer. In order to facilitate understanding of the solution, however, it will be solved by a hand calculation.

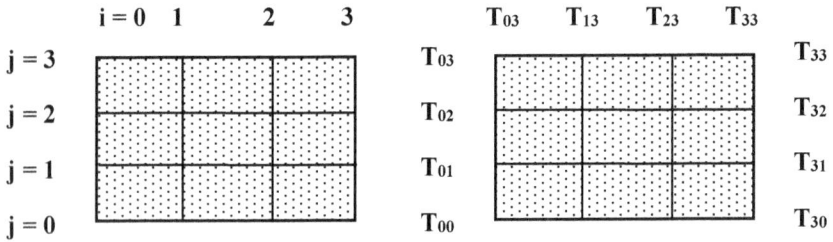

The complete presentation at every node is

	j = 1	$T_{11} = \dfrac{T_{01} + T_{21} + T_{10} + T_{12}}{4}$
i = 1	j = 2	$T_{12} = \dfrac{T_{02} + T_{22} + T_{11} + T_{13}}{4}$
	j = 3	$T_{13} = \dfrac{T_{03} + T_{23} + T_{12} + T_{14}}{4}$
	j = 1	$T_{21} = \dfrac{T_{11} + T_{31} + T_{20} + T_{22}}{4}$
i = 2	j = 2	$T_{22} = \dfrac{T_{12} + T_{32} + T_{21} + T_{23}}{4}$
	j = 3	$T_{23} = \dfrac{T_{13} + T_{33} + T_{22} + T_{24}}{4}$
	j = 1	$T_{31} = \dfrac{T_{21} + T_{41} + T_{30} + T_{32}}{4}$
i = 3	j = 2	$T_{32} = \dfrac{T_{22} + T_{42} + T_{30} + T_{32}}{4}$
	j = 3	$T_{33} = \dfrac{T_{23} + T_{43} + T_{32} + T_{34}}{4}$

T_{00}, T_{01}, T_{02} are each equal to 750 K. T_{13}, T_{23}, T_{33}, T_{32}, T_{31}, T_{30}, T_{10}, T_{20}, are all each, given, equal to 330 K. Note also that T_{41}, T_{42}, T_{43}, T_{14}, T_{24}, T_{34}, do not exist on the plate but may be assumed to be either zero or equal to 330K. It is safer to avoid the equations in which they occur as we have, in any case, four equations for the four unknowns we are looking for. The position of T_{03} is ambiguous in this problem since being on the T_{0j} face, it could be at 750 K while, also being at the T_{i3} face, it could also be at 330 K. Let us assume that it is 750 K. The difference equations that need to be solved become

$$T_{11} = \frac{750 + T_{21} + 330 + T_{12}}{4} \qquad (1)$$

$$T_{12} = \frac{750 + T_{22} + T_{11} + 330}{4} \qquad (2)$$

$$T_{21} = \frac{T_{11} + 330 + 330 + T_{22}}{4} \qquad (3)$$

$$T_{22} = \frac{T_{12} + 330 + T_{21} + 330}{4} \qquad (4)$$

We can solve equations (1) to (4) simultaneously as

$$4T_{11} - T_{12} - T_{21} = 1080 \qquad (5)$$

$$-T_{11} + 4T_{12} - T_{22} = 1080 \qquad (6)$$

$$-T_{11} + 4T_{21} - T_{22} = 660 \qquad (7)$$

$$-T_{12} - T_{21} + 4T_{22} = 1080 \qquad (8)$$

resulting in the matrix

$$\begin{pmatrix} 4 & -1 & -1 & 0 \\ -1 & 4 & 0 & -1 \\ -1 & 0 & 4 & -1 \\ 0 & -1 & -1 & 4 \end{pmatrix} \begin{pmatrix} T_{11} \\ T_{12} \\ T_{21} \\ T_{22} \end{pmatrix} = \begin{pmatrix} 1080 \\ 1080 \\ 660 \\ 660 \end{pmatrix} \qquad (9)$$

which has the solution

$$T_{11} = 487.5 \, K$$
$$T_{12} = 487.5 \, K$$
$$T_{21} = 382.5 \, K \qquad (10)$$
$$T_{22} = 382.5 \, K$$

The heat flux

$$q = k \frac{\Delta y}{\Delta x} \sum_{j=2}^{3} (T_{1j} - T_{2j}) = 39 \times 1 \times (T_{12} - T_{22} + T_{13} - T_{23})$$

$$= 39 \times (487.5 - 382.5 + 330 - 330)$$

$$= 4095 \ \frac{W}{m^2} \ per \ m \ of \ depth \quad Ans$$

Example 6.24: It is required to construct a plane heat source by implanting heating elements in a plane mild steel wall 0.10 m thick in the x direction and infinite in the y and z directions. If the outside surface is in contact with air at a temperature of 308 K with a heat transfer coefficient of 200 W/m²K while the inside surface is in contact with process material at 770 K with a heat transfer coefficient of 800 W/m²K, determine the temperature distribution in the plate if the heat release per unit volume is 15 kW/m³.

Answer: Suppose we divide the plane wall into 5 elements each of thickness 0.02m as shown below. There are two surface nodes, T_1 and T_6 and four interior nodes, T_2, T_3, T_4, and T_5.

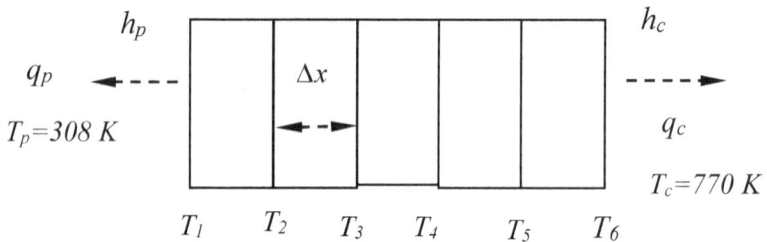

A heat balance method will be used to solve this problem, numerically. We shall assume, for simplicity, as this is, also, the common practical case, that the rate of heat release will be constant in all the nodes. Let us represent this heat release rate per unit area as U.

At steady state, the heat flow at any distance, y, at each node must balance the heat outflow from the node. Thus

At node 1	$q_{2 \to 1} - q_p + U = 0$	(1)
At node 2	$q_{1 \to 2} - q_{3 \to 2} + U = 0$	(2)
At node 3	$q_{2 \to 3} - q_{4 \to 3} + U = 0$	(3)
At node 4	$q_{3 \to 4} - q_{5 \to 4} + U = 0$	(4)
At node 5	$q_{4 \to 5} - q_{6 \to 5} + U = 0$	(5)
At node 6	$q_{5 \to 6} + q_c + U = 0$	(6)

When the heat flux terms are expressed in their standard forms, using numerical notation, we get

At node 1
$$\frac{k(T_2 - T_1)y}{\Delta x} - h_p(T_1 - T_p) + \frac{U\Delta x. y}{2} = 0 \qquad (7)$$

At node 2
$$\frac{k(T_1 - T_2)y}{\Delta x} - \frac{k(T_3 - T_2)y}{\Delta x} + U\Delta x. y = 0 \qquad (8)$$

At node 3
$$\frac{k(T_2 - T_3)y}{\Delta x} - \frac{k(T_4 - T_3)y}{\Delta x} + U\Delta x. y = 0 \qquad (9)$$

At node 4
$$\frac{k(T_3 - T_4)y}{\Delta x} - \frac{k(T_5 - T_4)y}{\Delta x} + U\Delta x. y = 0 \qquad (10)$$

At node 5
$$\frac{k(T_4 - T_5)y}{\Delta x} - \frac{k(T_6 - T_5)y}{\Delta x} + U\Delta x. y = 0 \qquad (11)$$

At node 6
$$\frac{k(T_5 - T_6)y}{\Delta x} - h_c(T_c - T_6) + \frac{U\Delta x. y}{2} = 0 \qquad (12)$$

Multiplying across by $\Delta x/y.k$, we get

At node 1
$$T_2 - T_1 - \frac{h_p \Delta x}{k}(T_1 - T_p) + \frac{U(\Delta x)^2}{2k} = 0 \qquad (13)$$

At node 2
$$T_1 - T_2 + T_3 - T_2 + \frac{U(\Delta x)^2}{k} = 0 \qquad (14)$$

At node 3
$$T_2 - T_3 + T_4 - T_3 + \frac{U(\Delta x)^2}{k} = 0 \qquad (15)$$

At node 4
$$T_3 - T_4 + T_5 - T_4 + \frac{U(\Delta x)^2}{k} = 0 \tag{16}$$

At node 5
$$T_4 - T_5 + T_6 - T_5 + \frac{U(\Delta x)^2}{k} = 0 \tag{17}$$

At node 6
$$T_5 - T_6 - \frac{h_c \Delta x}{k}(T_c - T_6) + \frac{U(\Delta x)^2}{2k} = 0 \tag{18}$$

These can be re-arranged as

$$-T_1\left(1 + \frac{h_p \Delta x}{k}\right) + T_2 = -T_p \frac{h_p \Delta x}{k} - \frac{U(\Delta x)^2}{2k} \tag{19}$$

$$T_1 - 2T_2 + T_3 = -\frac{U(\Delta x)^2}{k} \tag{20}$$

$$T_2 - 2T_3 + T_4 = -\frac{U(\Delta x)^2}{k} \tag{21}$$

$$T_3 - 2T_4 + T_5 = -\frac{U(\Delta x)^2}{k} \tag{22}$$

$$T_4 - 2T_5 + T_6 = -\frac{U(\Delta x)^2}{k} \tag{23}$$

$$T_5 - T_6\left(1 + \frac{h_c \Delta x}{k}\right) = -T_c \frac{h_c \Delta x}{k} - \frac{U(\Delta x)^2}{2k} \tag{24}$$

Since

$h_p = 200$ W/m²K; $h_c = 800$ W/m²K; $k = 39.6$ W/mK; $\Delta x = 0.02$ m $U = 15{,}000$ W/m³

$$T_p \frac{h_p \Delta x}{k} = 308 \, x \frac{200 \, x \, 0 \, 02}{39.6} , K. \frac{W}{m^2 K} . m . \frac{m.K}{W} = 31.1 \, K \tag{25}$$

$$1 + \frac{h_p \Delta x}{k} = 1 + \frac{200 \, x \, 0 \, 02}{39.6} . \frac{W}{m^2 K} . m . \frac{m.K}{W} = 1.1 \tag{26}$$

$$1 + \frac{h_c \Delta x}{k} = 1 + \frac{800 \, x \, 0 \, 02}{39.6} . \frac{W}{m^2 K} . m . \frac{m.K}{W} = 1.4 \tag{27}$$

$$T_c \frac{h_c \Delta x}{k} = 770\, x \frac{800\, x\, 0\, 02}{39.6}\, , K.\frac{W}{m^2 K}.m.\frac{m.K}{W} = 311.1\, K \quad (28)$$

$$\frac{U(\Delta x)^2}{2k} = \frac{15{,}000\, x\, (0.02)^2}{2\, x\, 39.6}\, \frac{W}{m^3}.m^2.\frac{m.K}{W} = 0.075\, K \quad (29)$$

$$\frac{U(\Delta x)^2}{k} = \frac{15{,}000\, x\, (0.02)^2}{39.6}\, \frac{W}{m^3}.m^2.\frac{m.K}{W} = 0.15\, K \quad (30)$$

Substituting these into equations (19) to (24), we have that

$$-1.1T_1 + T_2 = -31.175 \quad (31)$$
$$T_1 - 2T_2 + T_3 = -0.15 \quad (32)$$
$$T_2 - 2T_3 + T_4 = -0.15 \quad (33)$$
$$T_3 - 2T_4 + T_5 = -0.15 \quad (34)$$
$$T_4 - 2T_5 + T_6 = -0.15 \quad (35)$$
$$T_5 - 1.4T_6 = -311.175 \quad (36)$$

Equations (31) to (36) can, now, be arranged as the matrix

$$
\begin{pmatrix}
-1.1 & 1 & 0 & 0 & 0 & 0 \\
1 & -2 & 1 & 0 & 0 & 0 \\
0 & 1 & -2 & 1 & 0 & 0 \\
0 & 0 & 1 & -2 & 1 & 0 \\
0 & 0 & 0 & 1 & -2 & 1 \\
0 & 0 & 0 & 0 & 1 & -1.4
\end{pmatrix}
\begin{pmatrix}
T_1 \\ T_2 \\ T_3 \\ T_4 \\ T_5 \\ T_6
\end{pmatrix}
=
\begin{pmatrix}
-31.175 \\ -0.15 \\ -0.15 \\ -0.15 \\ -0.15 \\ -311.175
\end{pmatrix}
\quad (37)
$$

which has the $MATLAB^R$ solution

$$T_1 = 579.86\, K$$
$$T_2 = 606.67\, K$$
$$T_3 = 633.33\, K$$
$$T_4 = 659.84\, K$$
$$T_5 = 686.20\, K$$
$$T_6 = 712.41\, K \quad (38)$$

The matrix (37) is, also, a sparse, tri-diagonal matrix whose solution methods are detailed in standard mathematics text

References For Chapter Six

1   Carslaw, H.S and Jaeger, J.C., *Conduction of Heat in Solids*, Clarendon Press, Oxford, UK, 1959

2. Coulson J.M. and Richardson J.F. *Chemical Engineering Vol. 1* Pergamon Press, Oxford, UK, 1978

3   Heldman D.R: *Food Process Engineering*, Chapter 3, AVI Publishing Co; Westport, Connecticut, USA. 1975

4   Welty J.R: *Engineering Heat Transfer*, John Wiley and Sons New York, USA. 1978

5   Welty J.R; Wicks C.E and Wilson R.E: *Fundamentals of Momentum, Heat and Mass Transfer*; 2nd Edition; John Wiley and Sons, New York, USA, 1976.

6   Zill D. G. and Cullen M. R., *Differential Equations with Boundary Value Problems*; Brooks/Cole Publishing Co., California, USA, 1997

7. MATLAB; Version 7,0,4 365(R14) Service Pack 2, January 29, 2005; © 1994-2005 The MathWorks, Inc.

CHAPTER SEVEN
NON-STEADY STATE, ONE-DIMENSIONAL, HEAT CONDUCTION

Example 7.01: Why are we interested in non-steady state heat conduction?

Answer: Our interest arises from the reality of operating practical industrial and other processes in which these processes have to be started (start-ups), shut down (shut downs) or interrupted because of unexpected malfunctions or break down of steady state operations. Non-steady state analysis enables us understand and manage the behaviour of the process when not in steady state.

Example 7.02: Which of the heat conduction equations is applicable to non-steady state heat conduction?

Answer: The applicable general equation, in three dimensions, is the energy equation:

$$\rho Cp\frac{\partial T}{\partial t} = Q_V + k\left(\frac{\partial^2 T}{\partial x^2} + \frac{\partial^2 T}{\partial y^2} + \frac{\partial^2 T}{\partial z^2}\right) = Q_V + k\nabla^2 T \text{ (from 2.05)}$$

Example 7.03: What is a lumped parameter system in heat conduction in solids?

Answer: A lumped parameter system is that in which the heat transferred per unit time is solely dependent on the surface temperature distribution because its internal body temperature is uniform, that is, the surface resistance to heat transfer is dominant while its internal resistance to heat transfer is negligible.

Consider, for example, a body surrounded by a medium with which it exchanges heat energy. For such a body:

$Total\ resistance\ to\ heat\ transfer$
$$= Internal\ resistance + Surface\ resistance \quad (7.01)$$

If internal resistance is negligible, but surface resistance is not, the temperature within the body is independent of space co-ordinates or direction within that body, that is, the body is at uniform temperature. Temperature variations, if any, occur only at the surface of the body and with time. In other words

$$\nabla^2 T = \frac{\partial^2 T}{\partial x^2} + \frac{\partial^2 T}{\partial y^2} + \frac{\partial^2 T}{\partial z^2} = 0$$

and

$$\rho C_p \frac{\partial T}{\partial t} = Q_V \qquad (7.02)$$

The system described by equation (7.02) is called a lumped parameter system.

Example 7.04: What if a body in heat conduction is not a lumped parameter system, what does this mean?

Answer: When the system is not a lumped parameter system, that is, temperature varies within the body in spatial coordinates or direction, it means that the comparative values of internal and surface resistances influence the heat conduction. For example, if surface resistance is negligible, but internal resistance is not, then the surface temperature, T_S, is independent of spatial direction and may be assumed to be uniformly equal to the temperature of the surrounding medium, T_∞.

In this case, in which temperature variations occur within the body and also with time, the applicable, starting, equation is

$$\rho Cp \frac{\partial T}{\partial t} = Q_V + k\left(\frac{\partial^2 T}{\partial x^2} + \frac{\partial^2 T}{\partial y^2} + \frac{\partial^2 T}{\partial z^2}\right) = Q_V + k\nabla^2 T \quad \text{(from 2.05)}$$

In one dimension, this reduces to

$$\rho Cp \frac{\partial T}{\partial t} = Q_V + k\frac{\partial^2 T}{\partial x^2} \qquad (7.03)$$

In situations in which neither the internal nor the surface resistance to heat transfer can be considered negligible, both internal and surface resistance may be comparable in magnitude or in some proportion dependent on the particular system. Equation (7.03) will, still, apply.

Example 7.05: Derive the expression for heat conduction in a lumped parameter system without internal generation of heat.

Answer:

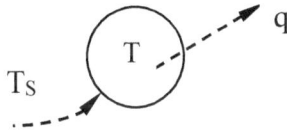

Consider such a body which is at a uniform internal temperature T. This body is in, and is surrounded by, an environment or medium at a temperature T_S. Suppose, as is often the case, that the initial and boundary conditions are:

$$t = 0 \quad T(0) = T_0$$
$$t > 0 \quad T(t) = T$$

Since no heat is absorbed or generated by the body $Q_V = 0$. Only external heating or cooling of the body is operative.

By a heat balance, since $Q_V = 0$:

153

$$Heat\ in = Heat\ out + Heat\ Accumulating \qquad (7.04)$$

In the situation in which the body is being cooled

$$0 = h_S A(T - T_\infty) + \rho V Cp \frac{dT}{dt} \qquad (7.04a)$$

When the body is being heated

$$h_S A(T_\infty - T) = 0 + \rho V Cp \frac{dT}{dt} \qquad (7.04b)$$

where h_S, is the surface convection coefficient and T_∞, is the temperature of the surroundings, far away from the surface of the body. A, ρ, V, Cp are, respectively, the surface area, density, volume and specific heat capacity of the body.

For the case in which the body is being cooled, equation (7.04a) can be rearranged as

$$\frac{dT}{dt} = \frac{h_S A}{\rho V Cp}(T_\infty - T)$$

which on integration

$$\int_{T_o}^{T} \frac{dT}{T_\infty - T} = \frac{h_S A}{\rho V Cp} \int_{0}^{t} dt$$

gives

$$\ln\left(\frac{T_\infty - T}{T_\infty - T_o}\right) = -\frac{h_S A}{\rho V Cp}t \qquad (7.05)$$

which can, also, be expressed as

$$\frac{T_\infty - T}{T_\infty - T_o} = e^{-\frac{h_S A}{\rho V Cp}t} \qquad (7.05a)$$

It can be seen that

$$\frac{h_S A}{\rho V Cp}t = \left(\frac{h_S V}{kA}\right)\left[\left(\frac{A}{V}\right)^2 \cdot \left(\frac{k}{\rho Cp}\right).t\right] = \left(\frac{h_S V}{kA}\right)\left(\frac{A^2 \alpha t}{V^2}\right) \qquad (7.05b)$$

154

$\frac{h_s V}{kA}$ is known as the Biot's number or modulus, *Bi*. It is dimensionless and represents the ratio of internal to surface resistance to heat transfer since

$$\frac{h_s V}{kA} = \frac{h_s}{kA/V} = \frac{h_s}{k/L} = \frac{L/k}{1/h_s} = \frac{internal\,resistance}{surface\,resistance} \qquad (7.05c)$$

$\frac{A^2 \alpha t}{V^2}$ is known as the Fourier number or modulus, Fo. It is, also, dimensionless and represents the ratio of real time, *t*, to the time defined by the properties of the body. That is

$$\frac{A^2 \alpha t}{V^2} = \frac{t}{V^2/A^2\alpha} = \frac{real\,time}{time\,defined\,by\,material\,property} (7.05d)$$

Note that V^2/A^2 has the units of (length)^2 while α, the thermal diffusivity, has units of (length)^2 per unit time. Hence equation (7.05a) can also be expressed as

$$\frac{T_\infty - T}{T_\infty - T_o} = e^{-Bi.Fo} \qquad (7.05e)$$

where *Bi* is the Biot number and *Fo* is the Fourier number. You may also find these numbers being represented as N_{Bi} and N_{Fo}.

Example 7.06: How do we know which resistance is dominant in heat conduction?

Answer: The above derivation and definition of the Biot's number enable us to develop criteria for identifying the operating regime of resistance to heat transfer during non-steady heat conduction. These criteria were developed from experience and are summarised below as

If Bi < 0.1	internal resistance is negligible
If Bi > 40	external resistance is negligible

> If $0.1 < Bi < 40$ internal and external resistance
> comparable

Example 7.07: A stainless steel billet, 150mm in diameter, is passing through a furnace 66 meters long. The initial billet temperature is 94°C and it must be raised to a minimum temperature of 816°C before working. The heat transfer coefficient between the furnace gases and the billet surface is 85W/m²K and the furnace gases are at 1260°C. At what minimum velocity must the billet travel through the furnace to satisfy these conditions? (Welty, 1978)

Answer: First compute

$$\frac{h_S V}{kA} \tag{1}$$

From Appendix H (Welty et al, 1976), for stainless steel

$\rho = 7820$ kg/m³   k $= 23$ W/m.K at 573 K

$Cp = 460.8$ J/kg K $\alpha = 0.44$ x 10^{-5} m²/s

Since

$$\frac{V}{A} = \frac{\pi D^3}{6} \cdot \frac{1}{\pi D^2} = \frac{D}{6} \tag{2}$$

$$Bi = \frac{h_S V}{kA} = \frac{85 \; x \; 0.15}{23 \; x \; 6} = 0.09 < 0.1 \tag{3}$$

That is; internal resistance is negligible.

From equation (7.05e)

$$\frac{T_\infty - T}{T_\infty - T_0} = \frac{1260 - 816}{1260 - 94} = \frac{444}{1166} = 0.3808 = e^{-0.09 \, .Fo}$$

That is

$$0.09 \cdot Fo = 0.9655 \quad or \quad Fo = 10.73 \qquad (4)$$

From (7.05d), (2) and (4)

$$Fo = 10.73 = \frac{A^2 \alpha\, t}{V^2} = \frac{36 \times 0.44 \times 10^{-5} \times t}{(0.15)^2} , \frac{m^4}{m^6} \cdot \frac{m^2}{s} \cdot s$$

That is

$$t = \frac{10.73 \times 0.0225}{36 \times 4.4 \times 10^{-6}} = 1524.2 \; s$$

Minimum Velocity

$$= \frac{66 \; m}{1524.2 \; s} = 0.0433 \frac{m}{s} \qquad Ans$$

Example 7.08: Derive the expression for heat conduction in a lumped parameter system with internal generation of heat.

Answer: When there is internal generation of heat during conduction, $Q_V \neq 0$, and the overall energy balance equation becomes

$$Heat\ in + Heat\ generated/consumed$$
$$= Heat\ out + Heat\ accumulated \qquad (7.06)$$

In the situation in which the body is being cooled

$$0 + Q_V = h_S A(T - T_\infty) + \rho V C p \frac{dT}{dt} \qquad (7.06a)$$

When the body is being heated

$$h_S A(T_\infty - T) + Q_V = 0 + \rho V C p \frac{dT}{dt} \qquad (7.06b)$$

where h_S, is the surface convection coefficient and T_∞, is the temperature of surroundings far away from the surface of the body. A, ρ, V, Cp are, respectively, the surface area, density, volume and specific heat capacity of the body. Note that ρV can be replaced by m, the mass of the body.

For the case in which the body is being cooled, such as in the use of the household electric iron, equation (4.6a) can be rearranged as

$$\frac{dT}{dt} = \frac{Q_V}{\rho V C p} - \frac{h_S A}{\rho V C p}(T - T_\infty)$$

$$= \frac{Q_V}{\rho V C p} - \frac{h_S A}{\rho V C p}T + \frac{h_S A}{\rho V C p}T_\infty \qquad (7.06c)$$

We can represent the coefficients and constants of the above equation as

$$a = \frac{Q_V}{\rho V C p} + \frac{h_S A}{\rho V C p}T_\infty \qquad (7.06d)$$

and

$$b = \frac{h_S A}{\rho V C p} \qquad (7.06e)$$

so that equation (4.6c) becomes

$$\frac{dT}{dt} = a - bT \qquad (7.06f)$$

which on integration

$$\int_{T_o}^{T} \frac{dT}{a - bT} = \int_{0}^{t} dt$$

gives

$$t = \left|-\frac{1}{b}\ln(a - bT)\right|_{T_o}^{T} = -\frac{1}{b}\ln(a - bT) + \frac{1}{b}\ln(a - bT_o)$$

which can, also, be expressed as

$$t = \frac{1}{b}\ln\left(\frac{a - bT_o}{a - bT}\right) \qquad (7.07)$$

Example 7.09: A household electric iron, rated at 500 W, has a stainless steel sole plate which weighs 1.36 kg and has a surface area of 0.05 m². The surroundings are at 27 C. The convective heat transfer coefficient between the sole plate and the surroundings is 17.00 W/m²K. Determine the time required for the iron to reach 116 C after it is plugged on and switched on. You are given that the Biot's number is

$$Bi = \frac{h_S V}{kA}$$

and the Fourier number is

$$Fo = \frac{A^2 \alpha t}{V^2}$$

where

h = convective heat transfer coefficient, W/m²K
V = volume of solid, m³
A = surface area of solid, m²
k = thermal conductivity of solid, W/m.K
α = thermal diffusivity of solid, m²/s
t = time, s
ρ = density of solid, kg/m³
Cp = specific heat capacity of solid at constant pressure, J/kg.K

For stainless steel, ρ = 7820 kg/m³, k = 7.6(1 + 3.42 x 10⁻³T) W/m.K, Cp = 4610 J/kg.K

Answer: Given h_S = 17.00 W/m²K; Cp = 4610 J/kg.K, A = 0.05 m², T_∞ = 27 C = 27 + 273 = 300 K, ρ = 7820 kg/m³, T = 116 C = 116 + 273 = 389 K

Since

$$k = 7.6(1 + 3.42 \; x \; 10^{-3} \; x \; 389) = 17.71 \; W/mK$$

and

$$V = \frac{m}{\rho} = \frac{1.36}{7820} \frac{kg}{kg/m^3} = 1.739 \; x \; 10^{-4} \; m^3$$

then

$$Bi = \frac{h_s V}{kA} = \frac{17 \; x \; 1.739 \; x \; 10^{-4}}{17.71 \; x \; 0.05} \; \frac{W}{m^2 K} . m^3 . \frac{mK}{W} . \frac{1}{m^2}$$

$$= 3.339 \; x \; 10^{-3} < 0.1$$

Hence we can assume that internal resistance to heat transfer is negligible. That is, only pure surface resistance exists. From the data given and equations (7.06d) and (7.06e)

$$a = \frac{Q_v}{\rho V C p} + \frac{h_s A}{\rho V C p} T_\infty$$

$$= \frac{500}{7820 \; x \; 1.739 \; x \; 10^{-4} \; x \; 4610} \frac{W}{1} \frac{m^3}{kg} \frac{1}{m^3} \frac{kg.K}{J}$$

$$+ \frac{17 \; x \; 0.05 \; x \; 300}{7820 \; x \; 1.739 \; x \; 10^{-4} \; x \; 4610} \frac{W}{m^2 K} \frac{m^2}{1} \frac{m^3}{kg} \frac{1}{m^3} \frac{kg.K}{J} . K$$

$$= 0.120 \frac{K}{s} \tag{1}$$

and

$$b = \frac{h_s A}{\rho V C p}$$

$$= \frac{17 \; x \; 0.05}{7820 \; x \; 1.739 \; x \; 10^{-4} \; x \; 4610} \frac{W}{m^2 K} \frac{m^2}{1} \frac{m^3}{kg} \frac{1}{m^3} \frac{kg.K}{J}$$

$$= 1.3559 \; x \; 10^{-4}, s^{-1} \tag{2}$$

Substituting (1) and (2) into equation (4.7) knowing that, in this case, $T_0 = T_\infty$, gives

$$t = \frac{1}{b} \ln\left(\frac{a - bT_o}{a - bT}\right)$$

$$= \frac{1}{1.3559 \; x \; 10^{-4}} \ln\left(\frac{0.120 - 1.3559 \; x \; 10^{-4} \; x \; 300}{0.120 - 1.3559 \; x \; 10^{-4} \; x \; 389}\right)$$

$$= 1217.1, s \quad Ans$$

Example 7.10: Outline the defining equations and the boundary and initial conditions for heat conduction in a system with negligible surface resistance.

Answer: This system is identified when Bi > 40. Usually, for all times greater than zero, that is $t > 0$, the surface temperature, T_S, is uniformly distributed, constant and equal to T_∞, the temperature of the surrounding environment. There are temperature variations within the system, however. The starting equation, applicable to temperature variations within the system, is

$$\rho C p \frac{\partial T}{\partial t} = Q_V + k\left(\frac{\partial^2 T}{\partial x^2} + \frac{\partial^2 T}{\partial y^2} + \frac{\partial^2 T}{\partial z^2}\right) = Q_V + k\nabla^2 T \; \text{(from 2.05)}$$

In one dimension, this reduces to

$$\rho C p \frac{\partial T}{\partial t} = Q_V + k\frac{\partial^2 T}{\partial x^2} \qquad (2.05a)$$

When there is no internal generation or consumption of heat energy, that is $Q_V = 0$, equation (2.05a) becomes

$$\rho C p \frac{\partial T}{\partial t} = k\frac{\partial^2 T}{\partial x^2} \qquad (7.08)$$

Equation (7.08) can be solved analytically by separation of variables, Laplace transformation or numerically by well-known methods.

For such a system in which there is no internal generation or consumption of heat energy, that is $Q_v = 0$, consider a solid slab of thickness, L, and, mathematically, of infinite length across which heat is being transferred as shown.

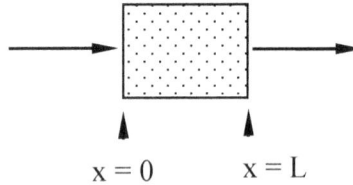

$$x = 0 \qquad x = L$$

Suppose the initial and boundary conditions are

(i) $t = 0;$ $0 \le x \le L;$ $T = T_0(x)$

(ii) $t > 0;$ $x = 0;$ $T = T_S$ (7.08a)

(iii) $t > 0;$ $x = L;$ $T = T_S$

The usual interest is to obtain a temperature distribution and its variation with time within the body as well as the heat flux for the heat transfer.

Example 7.11: How are these equations with their initial and boundary conditions solved?

Answer: These equations can be solved analytically using either separation of variables or a Laplace transform solution or by numerical methods.

Example 7.12: Outline the separation of variables solution.

Answer: It is usual, in this solution procedure, to normalise the temperature variable by defining

$$Y = \frac{T - T_S}{T_o - T_S} \qquad (7.09)$$

162

so that

$$dY = \frac{dT}{T_o - T_S} \qquad (7.09a)$$

The initial and boundary conditions become

(i) $t = 0$; $0 \le x \le L$; $Y = Y_0(x)$
(ii) $t > 0$; $x = 0$; $Y = 0$ $\qquad (7.09b)$
(iii) $t > 0$; $x = L$; $Y = 0$

If we apply the separation of variables technique, that

$$Y(x, t) = X(x) . \varphi(t) \qquad (7.10)$$

equation (7.08) becomes

$$X(x) \frac{\partial \varphi(t)}{\partial t} = \alpha. \varphi(t) \frac{\partial^2 X(x)}{\partial x^2} \qquad (7.11)$$

where

$$\alpha = \frac{k}{\rho C p} \qquad (7.11a)$$

To use the separation of variables technique we have to put equation (7.11) in the form

$$\frac{1}{\alpha. \varphi(t)} \frac{d\varphi(t)}{dt} = \frac{1}{X(x)} \frac{d^2 X(x)}{dx^2} = -\omega^2 = \text{constant} \qquad (7.11b)$$

That is

$$\frac{d\varphi(t)}{dt} + \alpha. \omega^2. \varphi(t) = 0 \qquad (7.11c)$$

$$\frac{d^2 X(x)}{dx^2} + \omega^2. X(x) = 0 \qquad (7.11d)$$

Equation (7.11c) has the solution

$$\varphi(t) = Ce^{-\alpha.\omega^2 t} \qquad (7.12)$$

163

while equation (7.11d) has the solution

$$X(x) = A\cos(\omega x) + B\sin(\omega x) \qquad (7.13)$$

where A, B and C are integration constants. From equations (7.10), (7.12) and (7.13)

$$Y(x,t) = [A\cos(\omega x) + B\sin(\omega x)].e^{-\alpha.\omega^2 t} \qquad (7.14)$$

where the new A = the old A x C and the new B = the old B x C.

Applying the boundary condition (ii) of (7.09b), that is at t > 0; x = 0; Y = 0

$$0 = [A\cos(0) + B\sin(0)].e^{-\alpha.\omega^2 t}$$

That is

$$A = 0 \qquad (7.14a)$$

Applying the boundary condition (iii) of (7.09b), that is at t > 0; x = L; Y = 0

$$0 = B\sin(\omega L)$$

Since B ≠ 0, then

$$\sin(\omega L) = 0 \quad or \quad \omega = \frac{n\pi}{L} \qquad (7.14b)$$

where n = 1, 2, 3, etc. Substituting (7.14a) and (7.14b) in (7.14), for all n

$$Y(x,t) = \sum_{n=1}^{\infty} e^{-\left(\frac{n\pi}{L}\right)^2 \alpha t} B_n \sin\left(\frac{n\pi}{L}x\right) \qquad (7.14c)$$

Applying the initial condition (i) of (7.09b), that is at t = 0; $0 \le x \le L$; Y = Y₀(x)

$$Y(x,0) = \sum_{n=1}^{\infty} B_n \sin\left(\frac{n\pi}{L}x\right) \qquad (7.14d)$$

164

This allows us, by Fourier series analysis technique, to obtain

$$B_n = \frac{2}{L} \int_0^L Y(x,0) \sin\left(\frac{n\pi}{L}x\right) dx \qquad (7.14e)$$

Substituting equation (7.14e) in equation (7.14c)

$$Y(x,t)$$
$$= \frac{2}{L} \sum_{n=1}^{\infty} e^{-\left(\frac{n\pi}{L}\right)^2 \alpha t} \sin\left(\frac{n\pi x}{L}\right) \int_0^L Y(x,0) \sin\left(\frac{n\pi x}{L}\right) dx \quad (7.14f)$$

For an infinite slab of initial uniform temperature, $Y_0(x) = Y_0 = 1$,

$$Y(x,t) = \frac{T - T_S}{T_o - T_S} = \frac{4}{\pi} \sum_{n=1}^{\infty} \frac{1}{n} e^{-\left(\frac{n\pi}{L}\right)^2 Fo} \sin\left(\frac{n\pi x}{L}\right) \qquad (7.14g)$$

where

$$Fo = \frac{\alpha t}{(L/2)^2}$$

and x is computed from the centre with n = 1,3,5, etc.

Similarly, it has been shown (Welty, 1978) that for an infinite cylinder, of initial uniform temperature, of radius, R, except at the centre, $Y_0(r) = Y_0 = 1$,

$$Y(x,t) = \frac{T - T_S}{T_o - T_S} = \sum_{n=1}^{\infty} \frac{2}{R_n} e^{-(R_n)^2 Fo} \frac{J_o\left(R_n \cdot \frac{r}{R}\right)}{J_1(R_n)} \qquad (7.14h)$$

where

$$Fo = \frac{\alpha t}{R^2}$$

R_n is function of the Biot number (hR/k) and J_0 and J_1 are Bessel's functions (obtainable from tables).
For a sphere (except at the centre) of radius R, of initial uniform

temperature, $Y_0(r) = Y_0 = 1$,

$$Y(x,t) = \frac{T - T_S}{T_o - T_S} = \frac{2}{\pi} \cdot \frac{R}{r} \sum_{n=1}^{\infty} \frac{(-1)^{n+1}}{n} e^{-(n\pi)^2 Fo} \sin\left(\frac{n\pi r}{R}\right) \quad (7.14i)$$

where

$$Fo = \frac{\alpha t}{R^2}$$

In industrial practice, these expressions are plotted in charts known, in the USA, as Heissler charts, which are plots of $\frac{T - T_S}{T_o - T_S}$ versus $\alpha t / R^2$ or $\alpha t / x^2$. .

Example 7.13: Outline the Laplace transform solution.

Answer: While separation of variable solutions are adequate for events that take place over long times, Laplace transform solutions are more suited to events that take place over very short times, such as the so called transients. Thus, while the boundary conditions for x, for separation of variables solution, may be 0 and L, for the Laplace transform solution, it would be 0 and ∞. The reason for this will soon be obvious. The equation to be solved is, still,

$$\rho Cp \frac{\partial T}{\partial t} = k \frac{\partial^2 T}{\partial x^2} \quad (7.08)$$

subject to the initial and boundary conditions

(i)	$t = 0$	$0 \le x \le L$	$T = T_0(x)$	
(ii)	$t > 0$	$x = 0$	$T = T_S$	(7.08a)
(iii)	$t > 0$	$x = L$	$T = T_S$	

We can, still, normalise the temperature variable by defining, as before,

$$Y = \frac{T - T_S}{T_o - T_S} \qquad (7.09)$$

so that

$$dY = \frac{dT}{T_o - T_S} \qquad (7.09a)$$

Equation (7.08) becomes

$$\frac{1}{T_o - T_S}\frac{\partial Y}{\partial t} = \frac{k}{\rho Cp} \cdot \frac{1}{T_o - T_S}\frac{\partial^2 Y}{\partial x^2}$$

or

$$\frac{\partial Y}{\partial t} = \alpha \frac{\partial^2 Y}{\partial x^2} \qquad (7.15)$$

where

$$\alpha = \frac{k}{\rho Cp}$$

The initial and boundary conditions become

(i) $t = 0$ $0 \le x \le L$ $Y = Y_0(x)$

(ii) $t > 0$ $x = 0$ $Y = 0$ (7.15a)

(iii) $t > 0$ $x = L$ $Y = 0$

If $\bar{Y}$ is the Laplace transform of Y, then, from equation (7.15)

$$\frac{\partial \bar{Y}}{\partial t} = \alpha \frac{\partial^2 \bar{Y}}{\partial x^2} \qquad (7.15b)$$

Since

$$\frac{\partial \bar{Y}}{\partial t} = p\bar{Y} - Y_o(x) \qquad (7.15c)$$

equation (7.15b) can be expressed as

$$p\bar{Y} - Y_o(x) = \alpha \frac{\partial^2 \bar{Y}}{\partial x^2}$$

167

That is

$$\frac{d^2\bar{Y}}{dx^2} - \frac{p}{\alpha}\bar{Y} + \frac{Y_0(x)}{\alpha} = 0 \qquad (7.15d)$$

For an initial uniform temperature, $Y_0(x) = Y_0$, equation (7.15d) becomes

$$\frac{d^2\bar{Y}}{dx^2} - \frac{p}{\alpha}\bar{Y} = -\frac{Y_0}{\alpha} \qquad (7.15e)$$

with the Laplace transformed boundary conditions

(i)	$t = 0$	$0 \le x \le L$	$\bar{Y} = \frac{Y_0}{\alpha}$	
(ii)	$t > 0$	$x = 0$	$\bar{Y} = 0$	(7.15f)
(iii)	$t > 0$	$x = L$	$\bar{Y} = 0$	

Equation (7.15e) has the solution

$$\bar{Y} = Ae^{\left(\sqrt{\frac{p}{\alpha}}\right)x} + Be^{-\left(\sqrt{\frac{p}{\alpha}}\right)x} + \frac{Y_0}{p} \qquad (7.15g)$$

Applying boundary condition (ii)

$$0 = A + B + \frac{Y_0}{p} \qquad (7.15h)$$

Applying boundary condition (iii)

$$0 = Ae^{\left(\sqrt{\frac{p}{\alpha}}\right)L} + Be^{-\left(\sqrt{\frac{p}{\alpha}}\right)L} + \frac{Y_0}{p} \qquad (7.15i)$$

Multiplying equation (7.15h) by $e^{\left(\sqrt{\frac{p}{\alpha}}\right)L}$ gives

$$0 = Ae^{\left(\sqrt{\frac{p}{\alpha}}\right)L} + Be^{\left(\sqrt{\frac{p}{\alpha}}\right)L} + \frac{Y_0}{p}e^{\left(\sqrt{\frac{p}{\alpha}}\right)L} \qquad (7.15j)$$

Subtracting equation (7.15i) from equation (7.15j), we get

$$B\left[e^{\left(\sqrt{\frac{p}{\alpha}}\right)L} - e^{-\left(\sqrt{\frac{p}{\alpha}}\right)L}\right] = \frac{Y_o}{p}\left[1 - e^{\left(\sqrt{\frac{p}{\alpha}}\right)L}\right]$$

or

$$B = \frac{\dfrac{Y_o}{p}\left[1 - e^{\left(\sqrt{\frac{p}{\alpha}}\right)L}\right]}{\left[e^{\left(\sqrt{\frac{p}{\alpha}}\right)L} - e^{-\left(\sqrt{\frac{p}{\alpha}}\right)L}\right]} \qquad (7.15k)$$

Similarly, multiplying equation (7.15h) by $e^{-\left(\sqrt{\frac{p}{\alpha}}\right)L}$ gives

$$0 = Ae^{-\left(\sqrt{\frac{p}{\alpha}}\right)L} + Be^{-\left(\sqrt{\frac{p}{\alpha}}\right)L} + \frac{Y_o}{p}e^{-\left(\sqrt{\frac{p}{\alpha}}\right)L} \qquad (7.15l)$$

Subtracting equation (7.15l) from equation (7.15i), we get

$$A\left[e^{\left(\sqrt{\frac{p}{\alpha}}\right)L} - e^{-\left(\sqrt{\frac{p}{\alpha}}\right)L}\right] = -\frac{Y_o}{p}\left[1 - e^{-\left(\sqrt{\frac{p}{\alpha}}\right)L}\right]$$

or

$$A = -\frac{\dfrac{Y_o}{p}\left[1 - e^{-\left(\sqrt{\frac{p}{\alpha}}\right)L}\right]}{\left[e^{\left(\sqrt{\frac{p}{\alpha}}\right)L} - e^{-\left(\sqrt{\frac{p}{\alpha}}\right)L}\right]} \qquad (7.15m)$$

Substituting for A and B in equation (7.15g)

$$\bar{Y} = -\frac{\frac{Y_0}{p}\left[1 - e^{-\left(\sqrt{\frac{p}{\alpha}}\right)L}\right]}{\left[e^{\left(\sqrt{\frac{p}{\alpha}}\right)L} - e^{-\left(\sqrt{\frac{p}{\alpha}}\right)L}\right]}e^{\left(\sqrt{\frac{p}{\alpha}}\right)x} + \frac{\frac{Y_0}{p}\left[1 - e^{\left(\sqrt{\frac{p}{\alpha}}\right)L}\right]}{\left[e^{\left(\sqrt{\frac{p}{\alpha}}\right)L} - e^{-\left(\sqrt{\frac{p}{\alpha}}\right)L}\right]}e^{-\left(\sqrt{\frac{p}{\alpha}}\right)x} + \frac{Y_0}{p}$$

$$\bar{Y} = \frac{-\frac{Y_0}{p}\left[e^{\left(\sqrt{\frac{p}{\alpha}}\right)x} - e^{-\left(\sqrt{\frac{p}{\alpha}}\right)L}e^{\left(\sqrt{\frac{p}{\alpha}}\right)x}\right] + \frac{Y_0}{p}\left[e^{-\left(\sqrt{\frac{p}{\alpha}}\right)x} - e^{\left(\sqrt{\frac{p}{\alpha}}\right)L}e^{-\left(\sqrt{\frac{p}{\alpha}}\right)x}\right]}{\left[e^{\left(\sqrt{\frac{p}{\alpha}}\right)L} - e^{-\left(\sqrt{\frac{p}{\alpha}}\right)L}\right]}$$
$$+ \frac{Y_0}{p}$$

$$\bar{Y} = \frac{-\frac{Y_0}{p}\left[e^{\left(\sqrt{\frac{p}{\alpha}}\right)x} - e^{-\left(\sqrt{\frac{p}{\alpha}}\right)x}\right] + \frac{Y_0}{p}\left[e^{-\left(\sqrt{\frac{p}{\alpha}}\right)(L-x)} - e^{\left(\sqrt{\frac{p}{\alpha}}\right)(L-x)}\right]}{\left[e^{\left(\sqrt{\frac{p}{\alpha}}\right)L} - e^{-\left(\sqrt{\frac{p}{\alpha}}\right)L}\right]} + \frac{Y_0}{p}$$

$$\bar{Y} = \frac{\frac{Y_0}{p}\left[e^{-\left(\sqrt{\frac{p}{\alpha}}\right)(L-x)} - e^{\left(\sqrt{\frac{p}{\alpha}}\right)(L-x)}\right]}{\left[e^{\left(\sqrt{\frac{p}{\alpha}}\right)L} - e^{-\left(\sqrt{\frac{p}{\alpha}}\right)L}\right]} - \frac{\frac{Y_0}{p}\left[e^{\left(\sqrt{\frac{p}{\alpha}}\right)x} - e^{-\left(\sqrt{\frac{p}{\alpha}}\right)x}\right]}{\left[e^{\left(\sqrt{\frac{p}{\alpha}}\right)L} - e^{-\left(\sqrt{\frac{p}{\alpha}}\right)L}\right]} + \frac{Y_0}{p} \qquad (7.15n)$$

The inverse Laplace transform of equation (7.15n) is not easily determined. If, however, in the boundary conditions (i) and (iii) of (7.15f), L is replaced by ∞, equation (7.15i) gives A = 0. Equation (7.15h) gives $B = -\frac{Y_0}{p}$ and equation (7.15g) becomes

$$\bar{Y} = \frac{Y_0}{p} - \frac{Y_0}{p}e^{-\left(\sqrt{\frac{p}{\alpha}}\right)x} \qquad (7.15o)$$

which gives the inverse Laplace transform,

$$Y = \frac{T - T_S}{T_0 - T_S} = Y_o - Y_o erfc\left(\frac{x}{2\sqrt{at}}\right) = erf\left(\frac{x}{2\sqrt{at}}\right) \qquad (7.16)$$

since $Y_0 = 1$,.

Example 7.14: Outline the numerical solutions.

Answer: The equation to be solved is, as before,

$$\rho Cp\frac{\partial T}{\partial t} = k\frac{\partial^2 T}{\partial x^2} \qquad (7.08)$$

subject to the initial and boundary conditions

(i)	t = 0	$0 \le x \le L$	$T = T_0(x)$
(ii)	t > 0	x = 0	$T = T_S$
(iii)	t > 0	x = L	$T = T_S$

(7.08a)

Equation (7.08) can, also, be expressed as

$$\frac{\partial T}{\partial t} = \alpha\frac{\partial^2 T}{\partial x^2} \qquad (7.08b)$$

where

$$\alpha = \frac{k}{\rho Cp}$$

At any given t, for an increment on x of h, the Taylor series approximation of $\frac{\partial^2 T}{\partial x^2}$ is

$$\left(\frac{\partial^2 T}{\partial x^2}\right)_i = \frac{T(x_i + h, t) + T(x_i - h, t) - 2T(x_i, t)}{h^2} \qquad (5.03)$$

At any given x, for an increment on t of k, the Taylor series approximation of $\frac{\partial T}{\partial t}$ is

$$\frac{\partial T}{\partial t} = \frac{T(x, t + k) - T(x, t)}{k} \qquad (7.17)$$

Thus, from equations (7.08b), (5.03) and (7.17)

$$\frac{T(x, t + k) - T(x, t)}{k} = \alpha \frac{T(x_i + h, t) + T(x_i - h, t) - 2T(x_i, t)}{h^2}$$

which can be simplified to

$$T(x, t + k) - T(x, t)$$

$$= \frac{\alpha k}{h^2} [T(x_i + h, t) + T(x_i - h, t) - 2T(x_i, t)] \qquad (7.18)$$

To simplify notations we can use the following simpler notations

Original Notation	Simplified Notation
$T(x, t + k)$	$T_{i, j+1}$
$T(x_i, t)$	$T_{i, j}$
$T(x_i + h, t)$	$T_{i+1, j}$
$T(x_i - h, t)$	$T_{i-1, j}$

Equation (7.18) becomes

$$T_{i,j+1} - T_{i,j} = \beta [T_{i+1,j} + T_{i-1,j} - 2T_{i,j}]$$

or

$$T_{i,j+1} = \beta T_{i+1,j} + (1 - 2\beta)T_{i,j} + \beta T_{i-1,j} \qquad (7.19)$$

where .

$$\beta = \frac{\alpha k}{h^2}$$

We can illustrate the two dimensional representation of the process as shown below (Zill & Cullen, 1997).

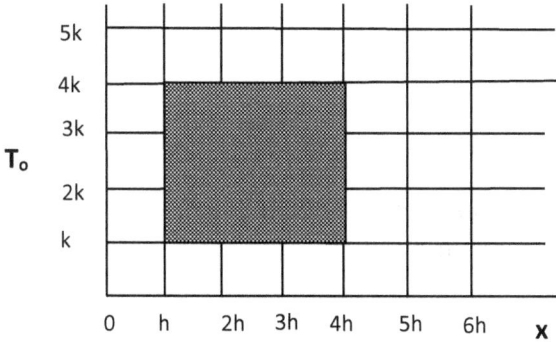

The nodes are shown as

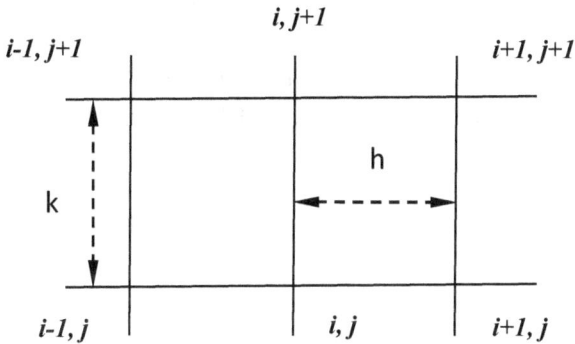

Thus, if the range of time is from θ_0 to θ and x ranges from x_1 to x_2, we can select h and k as we please such that

$$h = \frac{x_2 - x_1}{n} \quad and \quad k = \frac{\theta - \theta_o}{m} \qquad (7.20)$$

where n and m are the number of h and k units, respectively, we would like to work with.

Example 7.15: The numerical finite difference solution of the one dimensional unsteady state heat conduction equation,

173

$$\frac{\partial T}{\partial t} = \alpha \frac{\partial^2 T}{\partial x^2} \tag{7.08b}$$

is given by

$$T_{i,j+1} = \beta T_{i+1,j} + (1 - 2\beta)T_{i,j} + \beta T_{i-1,j} \tag{7.19}$$

where j = iteration number, i = nodal point, $\beta = \alpha k/h^2$, and h and k are increments on x and t.

A flat aluminium bar, 0.20m thick and sufficiently large for end effects to be neglected, is, initially, at a uniform temperature of 300K. At time t = 0 the top surface is suddenly heated to, and maintained at, 373K.

Determine the temperature distribution within the first 40 seconds. Take, for aluminium, that k = 230 W/m.K, ρ = 2700 kg/m³, Cp = 938 J/kg.K.

Answer: If we choose $n = 4$ and $m = 4$, then, from equation (7.20)

$$h = \frac{0.2 - 0}{4} = 0.05m \quad and \quad k = \frac{40 - 0}{4} = 10 \ s \quad (from \ 7.20)$$

$$\beta = \frac{\alpha k}{h^2} = \frac{230}{2700 \ x \ 938} x \frac{10}{(0.05)^2}, \frac{W}{mK} \cdot \frac{m^3}{kg} \cdot \frac{kgK}{J} \cdot \frac{s}{1} \cdot \frac{1}{m^2} = 0.363$$

Thus we get from equation (7.19)

$$T_{i,j+1} = 0.363T_{i+1,j} + 0.274T_{i,j} + 0.363T_{i-1,j} \quad (from \ 7.19)$$

For j = 0 (equivalent to initial condition, at t = 0), we can deduce, from the given initial and boundary conditions, the surface and initial conditions thus:

x/t	0	10	20	30	40
0	$T_{00} = 373$	$T_{01} = 373$	$T_{02} = 373$	$T_{03} = 373$	$T_{04} = 373$
0.05	$T_{10} = 300$	$T_{11} = 300$	$T_{12} = 300$	$T_{13} = 300$	$T_{14} = 300$
0.10	$T_{20} = 300$	$T_{21} = 300$	$T_{22} = 300$	$T_{23} = 300$	$T_{24} = 300$
0.15	$T_{30} = 300$	$T_{31} = 300$	$T_{32} = 300$	$T_{33} = 300$	$T_{34} = 300$
0.20	$T_{40} = 300$	$T_{41} = 300$	$T_{42} = 300$	$T_{43} = 300$	$T_{44} = 300$

For j = 1 (equivalent to t = 10 seconds),

$$T_{i,j+1} = 0.363T_{i+1,j} + 0.274T_{i,j} + 0.363T_{i-1,j} \qquad (7.19)$$

when i = 1 (x = 0.05m),

$$
\begin{aligned}
T_{12} &= 0.363T_{21} + 0.274T_{11} + 0.363T_{01} \\
&= 0.363 \ x \ 300 + 0.274 \ x \ 300 + 0.363 \ x \ 373 \\
&= 326.50
\end{aligned}
$$

when i = 2 (x = 0.10m),

$$
\begin{aligned}
T_{22} &= 0.363T_{31} + 0.274T_{21} + 0.363T_{11} \\
&= 0.363 \ x \ 300 + 0.274 \ x \ 300 + 0.363 \ x \ 300 \\
&= 300
\end{aligned}
$$

When i = 3 (x = 0.15m),

$$
\begin{aligned}
T_{32} &= 0.363T_{41} + 0.274T_{31} + 0.363T_{21} \\
&= 0.363 \ x \ 300 + 0.274 \ x \ 300 + 0.363 \ x \ 300 \\
&= 300
\end{aligned}
$$

When i = 4 (x = 0.20m),

$$
\begin{aligned}
T_{42} &= 0.363T_{51} + 0.274T_{41} + 0.363T_{31} \\
&= 0.363 \ x \ 300 + 0.274 \ x \ 300 + 0.363 \ x \ 300 = 300
\end{aligned}
$$

For j = 2 (equivalent to t = 20 seconds),

$$T_{i,j+1} = 0.363T_{i+1,j} + 0.274T_{i,j} + 0.363T_{i-1,j} \qquad (7.19)$$

when i = 1 (x = 0.05m),

$$T_{13} = 0.363T_{22} + 0.274T_{12} + 0.363T_{02}$$
$$= 0.363 \times 300 + 0.274 \times 326.50 + 0.363 \times 373$$
$$= 333.76$$

when i = 2 (x = 0.10m),

$$T_{23} = 0.363T_{32} + 0.274T_{22} + 0.363T_{12}$$
$$= 0.363 \times 300 + 0.274 \times 300 + 0.363 \times 326.50$$
$$= 309.62$$

When i = 3 (x = 0.15m),

$$T_{33} = 0.363T_{42} + 0.274T_{32} + 0.363T_{22}$$
$$= 0.363 \times 300 + 0.274 \times 300 + 0.363 \times 300$$
$$= 300$$

When i = 4 (x = 0.20m),

$$T_{43} = 0.363T_{52} + 0.274T_{42} + 0.363T_{32}$$
$$= 0.363 \times 300 + 0.274 \times 300 + 0.363 \times 300$$
$$= 300$$

For j = 3 (equivalent to t = 30 seconds)

$$T_{i,j+1} = 0.363T_{i+1,j} + 0.274T_{i,j} + 0.363T_{i-1,j} \qquad (7.19)$$

when i = 1 (x = 0.05m),

$$T_{14} = 0.363T_{23} + 0.274T_{13} + 0.363T_{03}$$
$$= 0.363 \times 309.62 + 0.274 \times 333.76 + 0.363 \times 373$$
$$= 339.24$$

when i = 2 (x = 0.10m),

$$T_{24} = 0.363T_{33} + 0.274T_{23} + 0.363T_{13}$$
$$= 0.363 \ x \ 300 + 0.274 \ x \ 309.62 + 0.363 \ x \ 333.76$$
$$= 314.89$$

When i = 3 (x = 0.15m),

$$T_{34} = 0.363T_{43} + 0.274T_{33} + 0.363T_{23}$$
$$= 0.363 \ x \ 300 + 0.274 \ x \ 300 + 0.363 \ x \ 309.62$$
$$= 303.49$$

When i = 4 (x = 0.20m),

$$T_{44} = 0.363T_{53} + 0.274T_{43} + 0.363T_{33}$$
$$= 0.363 \ x \ 300 + 0.274 \ x \ 300 + 0.363 \ x \ 300$$
$$= 300$$

For j = 4 (equivalent to t = 40 seconds), $T_{i,j+1} = T_{i,5}$ which is undefined. Thus, there is no need to continue the calculations. Thus, after 40 seconds, the temperature distribution is obtained as follows:

x/t	0	10	20	30	40
0	$T_{00}=373$	$T_{01}=373$	$T_{02}=373$	$T_{03}=373$	$T_{04}=373$
0.05	$T_{10}=300$	$T_{11}=300$	$T_{12}=326.50$	$T_{13}=333.76$	$T_{14}=339.24$
0.10	$T_{20}=300$	$T_{21}=300$	$T_{22}=300$	$T_{23}=309.62$	$T_{24}=314.89$
0.15	$T_{30}=300$	$T_{31}=300$	$T_{32}=300$	$T_{33}=300$	$T_{34}=303.49$
0.20	$T_{40}=300$	$T_{41}=300$	$T_{42}=300$	$T_{43}=300$	$T_{44}=300$

Example 7.16: There appear to be situations, in practice, in which the computed values, using equation (7.19) or similar, fluctuate between negative and positive values thus making the solution unstable. Is there a method which handles such situations?

Answer: The Crank-Nicholson implicit method (Zill & Cullen, 1997) is said to avoid this problem. Although we shall not get into the detailed use of this algorithm, it is necessary to describe its basic characteristics so that the interested student can pursue it further. In this algorithm, equation (5.03)

$$\left(\frac{d^2T}{dx^2}\right)_i = \frac{T(x_i + h) + T(x_i - h) - 2T(x_i)}{h^2} \tag{5.03}$$

is replaced by the average of two central difference quotients, one evaluated at t and the other at $t+k$. Thus

$$\left(\frac{d^2T}{dx^2}\right)_i = \frac{1}{2}\left[\frac{T(x_i + h, t) + T(x_i - h, t) - 2T(x_i, t)}{h^2}\right.$$

$$+\frac{1}{2}\left[\frac{T(x_i + h, t + k) + T(x_i - h,\ t + k) - 2T(x_i, t + k)}{h^2}\right] \tag{7.21}$$

Equation (7.08b) becomes

$$\frac{T(x_i, t + k) - T(x_i, t)}{k}$$

$$= \frac{\alpha}{2}\left[\frac{T(x_i + h, t) + T(x_i - h, t) - 2T(x_i, t)}{h^2}\right.$$

$$+\frac{\alpha}{2}\left[\frac{T(x_i + h, t + k) + T(x_i - h,\ t + k) - 2T(x_i, t + k)}{h^2}\right] \tag{7.22}$$

By letting $\beta = \frac{\alpha k}{h^2}$, equation (7.22) becomes

$$2T(x_i, t + k) - 2T(x_i, t)$$

$$= \beta[T(x_i + h, t) + T(x_i - h, t) - 2T(x_i, t)]$$

$$+\beta[T(x_i + h, t + k) + T(x_i - h,\ t + k) - 2T(x_i, t + k)]$$

This can be simplified to

$$2T(x_i, t + k) - \beta T(x_i + h, t + k) - \beta T(x_i - h, t + k) + 2\beta T(x_i, t + k)$$

$$= 2\beta T(x_i, t) + \beta T(x_i + h, t) + \beta T(x_i - h, t) - 2T(x_i, t)$$

That is

$$-\beta T(x_i - h, t + k) + 2(1 + \beta)T(x_i, t + k) - \beta T(x_i + h, t + k)$$

$$= \beta T(x_i - h, t) + 2(1 - \beta)T(x_i, t) + \beta T(x_i + h, t)$$

Dividing throughout by β

$$-T(x_i - h, t + k) + 2\left(\frac{1}{\beta} + 1\right)T(x_i, t + k) - T(x_i + h, t + k)$$

$$= T(x_i - h, t) + 2\left(\frac{1}{\beta} - 1\right)T(x_i, t) + T(x_i + h, t) \qquad (7.23)$$

To simplify notation further, let us revert to *i,j* subscripts and define

$$\theta = 1 + \frac{1}{\beta} \qquad (7.24)$$

$$\gamma = 1 - \frac{1}{\beta} \qquad (7.25)$$

Equation (7.23) becomes

$$-T_{i-1,j+1} + 2\theta T_{i,j+1} - T_{i+1,j+1} = T_{i-1,j} - 2\gamma T_{i,j} + T_{i+1,j} \qquad (7.26)$$

For a given set of initial and boundary conditions, equation (7.26) can be developed into a matrix which can then be solved.

Example 7.17: Outline the solution procedures for heat conduction in systems with comparable surface and internal resistance

Answer: In this case, $0.1 < \text{Bi} < 40$ and solutions are often based on solutions obtained, previously, for either negligible internal resistance or negligible surface resistance, but applied along the centre or plane of symmetry of the object. It is usual to take this centre or plane of symmetry as the origin.

179

For example, for the infinite slab, the line of symmetry divides the infinite slab into two symmetrical halves, each of thickness L.

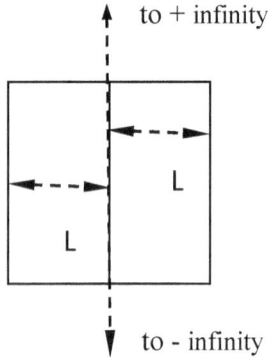

For the case in which there is no internal generation of heat, the equation to be solved is, as before,

$$\rho Cp \frac{\partial T}{\partial t} = k \frac{\partial^2 T}{\partial x^2} \qquad (7.08)$$

subject to the initial and boundary conditions

(i) $t = 0$ $0 \le x \le L$ $T = T_0$, uniform

(ii) $t > 0$ $x = 0$, centreline $\dfrac{\partial T(0, t)}{\partial x} = 0$

(iii) $t > 0$ $x = \pm L$ $-k \dfrac{\partial T(L, t)}{\partial x} =$ (7.27)

$$h[T(L, t) - T_\infty]$$

A separation of variables solution, with $Y = \dfrac{T - T_S}{T_0 - T_S}$, gives

$$Y(x, t) = e^{-\alpha \omega^2 t}[A cos (\omega x) + B sin (\omega x)] \qquad (7.28)$$

Applying boundary condition (ii)

$$0 = e^{-\alpha\omega^2 t}[-A\sin(0) + B\cos(0)]$$

shows that B = 0 so that

$$Y(x,t) = e^{-\alpha\omega^2 t}[A\cos(\omega x)] \tag{7.29}$$

Applying boundary condition (iii), at $x = \pm L$

$$\frac{\partial T(L,t)}{\partial x} = hY(L,t) \tag{7.30}$$

Hence at x = L, from (7.29) and (7.30)

$$k\omega A e^{-\alpha\omega^2 t}\sin(\omega x) = h e^{-\alpha\omega^2 t} A\cos(\omega x)$$

That is

$$\tan(\omega x) = \frac{h}{\omega k} = \tan(\omega L) = \frac{hL}{\omega kL} = \frac{Bi}{\omega L} \tag{7.31}$$

The general solution of equation (7.29) is of the form

$$Y(x,t) = \sum_{n=1}^{\infty} A_n e^{-\alpha\omega_n^2 t}\cos(\omega_n x) \tag{7.32}$$

At t = 0, equation (7.32) becomes

$$Y_0 = \sum_{n=1}^{\infty} A_n \cos(\omega_n x) \tag{7.33}$$

Multiplying both sides of equation (7.33) by $\cos(\omega_m x)$ and integrating, we get that

$$Y_0 \int_0^L \cos(\omega_n x)\, dx = \int_0^L \sum_{n=1}^{\infty} A_n \cos(\omega_n x)\cos(\omega_m x)\, dx$$

The right hand side of this equation is zero when $m \neq n$. For only n values, therefore,

$$A_n = Y_o \frac{\int_0^L \cos(\omega_n x)\, dx}{\int_0^L \cos^2(\omega_n x)\, dx} = Y_o \frac{2\sin(\omega_n L)}{\omega_n L + \sin(\omega_n L)\cos(\omega_n L)} \qquad (7.34)$$

Substituting equation (7.34) into equation (7.32)

$$\frac{Y(x,t)}{Y_o} = 2\sum_{n=1}^{\infty}\left[\frac{e^{-\alpha\omega_n^2 t}\sin(\omega_n L)\cos(\omega_n x)}{\omega_n L + \sin(\omega_n L)\cos(\omega_n L)}\right] \qquad (7.35)$$

This can be rearranged as

$$\frac{Y(x,t)}{Y_o} = 2\sum_{n=1}^{\infty}\left[\frac{e^{-\delta_n^2 Fo}\sin(\delta_n)\cos\left(\frac{\delta_n x}{L}\right)}{\delta_n + \sin(\delta_n)\cos(\delta_n)}\right] \qquad (7.36)$$

where
$$\delta_n = \omega_n L \qquad (7.36a)$$

and
$$Fo = \frac{\alpha t}{L^2} \qquad (7.36b)$$

and
$$\delta_n \tan \delta_n = Bi \qquad (7.36c)$$

Hence the temperature at any point, (x, t), is a function of x/L and of the Biot's (Bi) and Fourier's (Fo) numbers.

Similar expressions have been developed for cylinders and spheres. For an infinite cylinder, for example, (Welty, 1978)

$$\frac{Y(r,t)}{Y_o} = 2\sum_{n=1}^{\infty}\left[\frac{e^{-\delta_n^2 Fo}}{\delta_n} \cdot \frac{J_1(\delta_n)\cdot J_0\left(\frac{\delta_n r}{R}\right)}{J_0^2(\delta_n) + J_1^2(\delta_n)}\right] \qquad (7.37)$$

where
$$\frac{\delta_n J_1(\delta_n)}{J_0(\delta_n)} = \frac{hR}{k} \qquad (7.37a)$$

$J_0(\delta_n)$ and $J_1(\delta_n)$ are Bessel's functions of the first kind of zero and first orders, respectively.

For a sphere (Welty, 1978)

$$\frac{Y(r,t)}{Y_0} = 4\frac{R}{r}\sum_{n=1}^{\infty}\left[e^{-\delta_n^2 Fo}\sin(\delta_n)\cdot\frac{r}{R}\cdot\frac{\sin\delta_n - \delta_n\cos\delta_n}{2\delta_n - \sin(2\delta_n)}\right] \quad (7.38)$$

where

$$1 - \delta_n\cot\delta_n = \frac{hR}{k} \quad (7.38a)$$

Example 7.18: How are these equations used in practice?

Answer: Because industrial and in plant, even personal individual, practice prefers charts and quick and easily accessible values of desired operating or control parameters, it has been found useful to employ standardised plots of these equations in charts where the following terms are defined as listed below.

Y (dimensionless temperature) $\quad Y = \dfrac{T - T_\infty}{T_0 - T_\infty}$

X (relative time) $\quad X = Fo = \dfrac{\alpha t}{L^2}$

n (relative position) $\quad n = \dfrac{X}{L}$

m (relative resistance) $\quad m = \dfrac{k}{hL} = \dfrac{1}{Bi}$

Three types of charts are in use namely

(a). Temperature distribution charts for spheres, infinite slabs,

183

and cylinders.
(b) Energy efficiency charts (Q/Q_{inf}) for spheres, infinite slabs, and cylinders.
(c) Centre temperature charts for spheres and infinite cylinders.

Some of these are shown in Appendix I.

Example 7.19: A cylindrical mild steel slab 0.4m diameter and 44mm thick is exposed to air at 705 K. The surface coefficient is 17 W/m²K. If the mild steel was initially at a uniform temperature of 295 K, how long will the steel have to be treated for its temperature everywhere to be 660 K.

For mild steel, k = 39.6 W/mK, ρ = 7820 kg/m³, Cp = 473 J/kgK

Take the Biot number $= \dfrac{hV}{kA}$ and the Fourier number $= \dfrac{A^2 \alpha t}{V^2}$
where the symbols have their usual meaning in this area of study of heat transfer.

Answer: The slab may be represented, schematically, as shown. The surface area, A, the volume of the slab, V, the Biot's and Fourier's numbers may, also, be calculated as follows

0.4 m

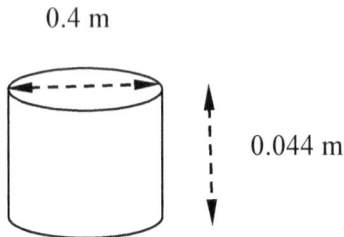

0.044 m

$$A = 2\left(\frac{\pi D^2}{4}\right) + \pi D L = \pi\left[\frac{D^2}{2} + DL\right] = \pi\left[\frac{(0.4)^2}{2} + 0.044 \times 0.4\right]$$

$$= 0.3067 \ m^2 \tag{1}$$

$$V = \frac{\pi D^2 L}{4} = \frac{\pi \times (0.4)^2 \times 0.044}{4} = 0.00553 \ m^3 \tag{2}$$

$$Bi = \frac{hV}{kA} = \frac{17 \times 0\ 00553}{39.6 \times 0.3067} \cdot \frac{W}{m^2 K} \cdot \frac{m^3}{1} \cdot \frac{mK}{W} \cdot \frac{1}{m^2} = 0.00774 \ < 0.1 \ (3)$$

Hence the system is a lumped parameter system. Thus

$$\frac{T - T_\infty}{T_o - T_\infty} = e^{-Bi.Fo} \tag{4}$$

But

$$\frac{T - T_\infty}{T_o - T_\infty} = \frac{660 - 705}{295 - 705} = 0.1098 \tag{5}$$

and

$$\alpha = \frac{k}{\rho Cp} = \frac{39.6}{7820 \times 473} \cdot \frac{W}{mK} \cdot \frac{m^3}{kg} \cdot \frac{kgK}{J} = 1.0706 \ \times 10^{-5} \ (6)$$

$$Fo = \frac{A^2 \alpha t}{V^2} = \frac{(0.3067)^2 \times 1.0706 \times 10^{-5} \times t}{(0.00553)^2} \cdot \frac{m^4}{m^6} \cdot \frac{m^2}{s} \cdot s$$

$$= 0.0329t \tag{7}$$

Then from (3), (4), (5) and (7)

$$\frac{T - T_\infty}{T_o - T_\infty} = 0.1098 = e^{-0.00774 \times 0.0329t}$$

That is

$$\ln(0.1098) = -0.00774 \times 0.0329t = 0.000254t = -2.2091 \text{or}$$

or

$$t = 8673.3 \ s = 2.41 \ hrs. \quad Ans$$

Example 7.20: The unsteady state temperature difference between any point in either half of an infinite bar is given as

$$\frac{\theta(x,t)}{\theta_o} = 2 \sum_{n=1}^{\infty} \left[\frac{e^{-\delta_n^2 Fo} \sin(\delta_n) \cos\left(\frac{\delta_n x}{L}\right)}{\delta_n + \sin(\delta_n) \cos(\delta_n)} \right] \qquad (7.36)$$

where

 x = any distance from the axis of symmetry
 L = thickness of the bar from the axis of symmetry
 $\delta_n \tan(\delta_n)$ = Bi, the Biot Number
 $\theta (x,t) = T(x,t) - T_\infty$
 $\theta_0 = T(x,0) - T_\infty$
 T_∞ = temperature at t > 0 and $x \geq \pm L$
 T(x, t) = temperature at t > 0 and at any $x \leq \pm L$
 T(x, 0) = temperature at t = 0 at and at any $x \leq \pm L$

If t = 30 seconds, T_∞ = 30 C, T(x,0) = 100 C, α = 1.54 x $10^{-5} m^2/s$ determine T(x, t) at a distance x = 3.2 cm from the centre of a slab 8 cm thick for which Bi = 0.4.

Answer: The solution of equation (7.36) is possible only when the equation (7.36c) is solved. This equation may be rearranged as

$$\tan \delta_n = \frac{Bi}{\delta_n} = \frac{0.4}{\delta_n} \qquad (1)$$

By letting

$$Y_1 = \tan \delta_n \quad and \quad Y_2 = \frac{0.4}{\delta_n} \qquad (2)$$

we can use an iterative algorithm or plot Y_1 versus δ_n and Y_2 versus δ_n of equation (2) to get a solution of equation (1) wherever $Y_1 = Y_2$. Such solutions will give us the values of δ_1, δ_2, δ_3, etc. Note that $\tan \delta_n$ is a trigonometric function while $\frac{0.4}{\delta_n}$ is a hyperbola. Since δ_n is now the independent variable, we can select any values we like. A manual, rather than a computer algorithm, solution, makes things clearer. Thus

Table 1 (Example 7.20)

δ_n, radians	$Y_1 = \tan \delta_n$	$Y_2 = \dfrac{0.4}{\delta_n}$
0	0.0000	
$\pi/8$	0.4142	1.0186
$\pi/4$	1.0000	0.5093
$3\pi/8$	2.4142	0.3395
$\pi/2$	∞	0.2546
$5\pi/8$	-2.4142	0.2037
$3\pi/4$	-1.0000	0.1698
$7\pi/8$	-0.4142	0.1455
π	0.0000	0.1273
$9\pi/8$	0.4142	0.1132
$5\pi/4$	1.0000	0.1019
$11\pi/8$	2.4142	0.0926
$3\pi/2$	∞	0.0849
$13\pi/8$	-2.4142	0.0784
$7\pi/4$	-1.0000	0.0728
2π	0.0000	0.0637

These are plotted in Fig. 1 (Example 7.20) below and give the solutions

$$\delta_1 = 0.589 \,(0.1875\pi)$$

$$\delta_2 = 3.5117 \,(1.1178\pi)$$

$$\delta_3 = 6.5298 \,(2.0785\pi)$$

Fig. 1: (Example 7.20)

Plot of $Y_1 = \tan \delta$ and $Y_2 = 0.4/\delta$ versus δ

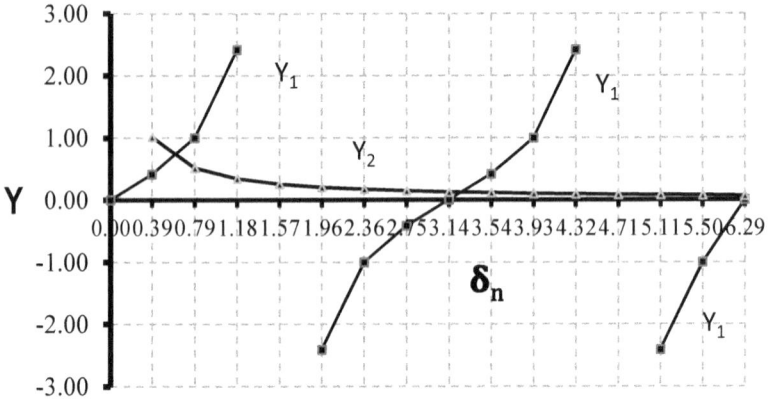

Values of δ_4, δ_5 and higher can be seen, from the graph, to give negligible values of Y_1 and Y_2. Equation (7.36) can now be evaluated. Still using a hand calculation, it is easier to proceed in a tabular and systematic form as follows. Since

$$\frac{x}{L} = \frac{3.2}{4.0} = 0.8 \tag{3}$$

and

$$\frac{\alpha t}{L^2} = \frac{1.54 \, x \, 10^{-5} \, x \, 30}{(0.04)^2} \frac{m^2}{s} \frac{s}{m^2} = 0.28875 \tag{4}$$

the table can be generated as shown in Table 2 of Example 7.20.

188

Table 2 (Example 7.20)

	δ_1 = 0.589	δ_2 = 3.5117	δ_3 = 6.5298
$\delta_n^2 \cdot \dfrac{\alpha t}{L^2}$	0.10017	3.5608	12.3118
$e^{-\delta_n^2 \cdot \frac{\alpha t}{L^2}}$	0.9047	0.0284	0
$\sin \delta_n$	0.5555	-0.3617	0.2441
$\cos \dfrac{\delta_n x}{L}$	0.8910	-0.9453	0.4894
$\sin \delta_n \cos \dfrac{\delta_n x}{L}$	0.4950	0.3419	0.1195
$\cos \delta_n$	0.8315	-0.9323	0.9697
$\sin \delta_n \cos \delta_n$	0.4619	0.3372	0.2367
$\delta_n + \sin \delta_n \cos \delta_n$	1.0509	3.8489	6.7665
$\dfrac{\sin \delta_n \cos \frac{\delta_n x}{L}}{\delta_n + \sin \delta_n \cos \delta_n}$	0.4710	0.0888	0.0177
$\dfrac{e^{-\delta_n^2 \cdot \frac{\alpha t}{L^2}} \sin \delta_n \cos \frac{\delta_n x}{L}}{\delta_n + \sin \delta_n \cos \delta_n}$	0.4261	0.0025	0

From this table

$$\frac{\theta(x,t)}{\theta_o} = 2 \sum_{n=1}^{\infty} \left[\frac{e^{-\delta_n^2 F_o} \sin(\delta_n) \cos\left(\frac{\delta_n x}{L}\right)}{\delta_n + \sin(\delta_n) \cos(\delta_n)} \right] = 0.4261 + 0.0025 + 0$$

$$= 2 \times 0.4286 = 0.8572 \qquad (5)$$

Since T(3.2, 0) = 100 C and T_∞ = 30 C, θ_0 = (100+273) – (30+273) = 70 K then, from (5)

$$\theta(x,t) = 0.8572\theta_o = T(3.2, 30) - (30 + 273)$$

$$= 0.8572 \times 70 = 60.00$$

Hence

$$(3.2, 30) = 303 + 60.00 = 363\ K\ or\ 90\ C\ Ans.$$

Example 7.21: Suppose the infinite bar of Example 7.20 was an infinite rod of 8 cm diameter, what will be the temperature at a radius of 3.2 cm? Take the surface coefficient, h, to be equal to 1024 W/m²K, the thermal diffusivity 1.54 x 10⁻⁵ m²/s, and the thermal conductivity to be 51.2 W/mK.

Answer: The appropriate equation for an infinite cylinder is equation (7.37)

$$\frac{Y(r,t)}{Y_o} = 2 \sum_{n=1}^{\infty} \left[\frac{e^{-\delta_n^2 F_o}}{\delta_n} \cdot \frac{J_1(\delta_n) \cdot J_0\left(\frac{\delta_n r}{R}\right)}{J_0^2(\delta_n) + J_1^2(\delta_n)} \right] \qquad (7.37)$$

where

$$\frac{\delta_n J_1(\delta_n)}{J_0(\delta_n)} = \frac{hR}{k} \qquad (7.37a)$$

and $J_0(\delta_n)$ and $J_1(\delta_n)$ are Bessel's functions of the first kind of zero and first orders, respectively.

The values of $J_0(\delta_n)$ and $J_1(\delta_n)$ are given in Korn & Korn (1968) as

$$J_0(x) = 1 - \frac{x^2}{2^2(1!)^2} + \frac{x^4}{2^4(2!)^2} - \frac{x^6}{2^6(3!)^2} + \cdots$$

$$J_1(x) = \frac{x}{2.0!.1!} - \frac{x^3}{2^3.1!.2!} + \frac{x^5}{2^5.2!.3!} - \cdots$$

As in Example 7.20, equation (7.37) cannot be solved without solving equation (7.37a) for values of $\delta_1, \delta_2, \delta_3,$, etc. Thus

$$\frac{\delta_n J_1(\delta_n)}{J_0(\delta_n)} = \frac{hR}{k} = \frac{1024 \times 0.04}{51.2}\ \frac{W}{m^2 K} \cdot m \cdot \frac{mK}{W} = 0.8$$

That is

190

$$\frac{J_1(\delta_n)}{J_0(\delta_n)} = \frac{0.8}{\delta_n} \tag{1}$$

As before, let

$$Y_1 = \frac{J_1(\delta_n)}{J_0(\delta_n)} \quad and \quad Y_2 = \frac{0.8}{\delta_n} \tag{2}$$

A plot of Y_1 versus δ_n and Y_2 versus δ_n will solve the equation whenever $Y_1 = Y_2$. Thus, using values of δ_n from Example (7.20) and values of $J_0(\delta_n)$ and $J_1(\delta_n)$ from tables in Korn & Korn, (1968), we have

Table 1 (Example 7.21)

δ_n	$J_0(\delta_n)$	$J_1(\delta_n)$	$Y_1 = \dfrac{J_1(\delta_n)}{J_0(\delta_n)}$	$Y_2 = \dfrac{0.8}{\delta_n}$
0.0	1.0000	0.0000	0.000	∞
0.4	0.9604	0.1960	0.204	2.000
0.8	0.8463	0.3688	0.436	1.000
1.2	0.6711	0.4983	0.743	0.667
1.6	0.4554	0.5699	1.251	0.500
2.0	0.2239	0.5767	2.576	0.400
2.4	0.0025	0.5202	208.08	0.333
2.8	-0.1850	0.4097	-2.215	0.286
3.2	-0.3202	0.2613	-0.816	0.250

3.6	-0.3918	0.0955	-0.244	0.222
4.0	-0.3971	-0.0660	0.166	0.200
4.4	-0.3423	-0.2028	0.593	0.182
4.8	-0.2404	-0.2985	1.242	0.167
5.2	-0.1103	-0.3432	3.112	0.154
5.6	0.0270	-0.3343	-12.382	0.143
6.0	0.1506	-0.2767	-1.837	0.133
6.4	0.2433	-0.1816	-0.746	0.125
6.8	0.2931	-0.0652	-0.222	0.118
7.2	0.2951	0.0543	0.184	0.111
7.6	0.2516	0.1592	0.633	0.105

Fig. 1 (Example 7.21)

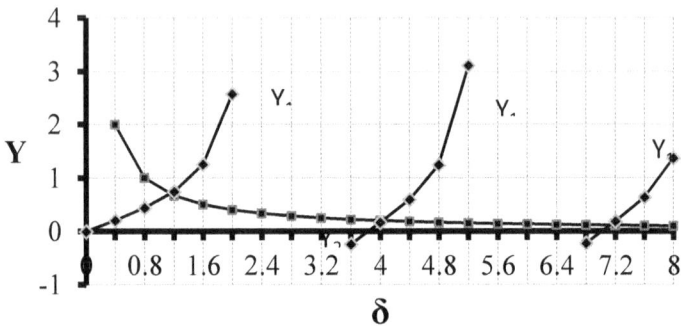

These are plotted in the figure above (positive values only) and give the solutions $\delta_1 = 1.15$, $\delta_2 = 4.05$, $\delta_3 = 7.15$.

To evaluate equation (7.37), calculate

$$\frac{r}{R} = \frac{3.2}{4.0} = 0.8 \qquad (3)$$

and

$$\frac{\alpha t}{R^2} = \frac{1.54 \times 10^{-5} \times 30}{(0.04)^2} \frac{m^2}{s} \frac{s}{m^2} = 0.28875 \qquad (4)$$

A table, convenient for a hand calculation, can be generated as shown below.

Table 2 (Example 7.21)

	$\delta_1 = 1.15$	$\delta_2 = 4.05$	$\delta_3 = 7.15$
$\delta_n{}^2 \dfrac{\alpha t}{R^2}$	0.3819	4.7362	14.7616
$\dfrac{e^{-\delta_n{}^2 \frac{\alpha t}{R^2}}}{\delta_n}$	0.5935	0.0022	0
$\dfrac{\delta_n r}{R}$	0.3321	1.1694	2.0646
$J_0\left(\dfrac{\delta_n r}{R}\right)$	0.9730	0.6860	0.1837
$J_1(\delta_n)$	0.4850	- 0.0849	0.0398
$J_1(\delta_n) \cdot J_0\left(\dfrac{\delta_n r}{R}\right)$	0.4719	- 0.0582	0.0073
$J_0(\delta_n)$	0.6957	- 0.3934	0.2974

$J_0^2(\delta_n) + J_1^2(\delta_n)$	0.7192	0.1620	0.0900
$\dfrac{J_1(\delta_n) \cdot J_0\left(\frac{\delta_n r}{R}\right)}{J_0^2(\delta_n) + J_1^2(\delta_n)}$	0.6562	0.3593	0.0811
$e^{-\delta_n^2 F_o} \dfrac{J_1(\delta_n) \cdot J_0\left(\frac{\delta_n r}{R}\right)}{\delta_n \cdot J_0^2(\delta_n) + J_1^2(\delta_n)}$	0.3895	- 0.0008	0

From this table and equation (7.37)

$$\frac{Y(3.2, 30)}{Y_o} = 2 \sum_{n=1}^{\infty} \left[e^{-\delta_n^2 F_o} \frac{J_1(\delta_n) \cdot J_0\left(\frac{\delta_n r}{R}\right)}{\delta_n \cdot J_0^2(\delta_n) + J_1^2(\delta_n)} \right]$$

$$= 2[0.3895 - 0.0008 + 0] = 0.7774$$

That is

$$\frac{Y(3.2, 30)}{Y_o} = 0.7774 = \frac{T(3.2, 30) - (30 + 273)}{(100 + 273) - (30 + 273)}$$

From which

$$T(3.2, 30) = 303 + 0.7774 \ x \ 70 = 357.42 \ K \ or \ 84.42 \ C. \ Ans$$

Example 7.22: The solution to the one dimensional equation for transient heat conduction in a semi-infinite medium

$$\frac{1}{\alpha} \frac{\partial T}{\partial t} = \frac{\partial^2 T}{\partial x^2}$$

is given by

$$\frac{T - T_\infty}{T_S - T_\infty} = 1 - erf\left(\frac{x}{2\sqrt{\alpha t}}\right)$$

when the initial and boundary conditions are described by

$$t = 0 \quad x > 0 \qquad\qquad T = T_\infty$$

194

$$t > 0 \quad x = 0 \text{ (surface of wall)} \quad T = T_S$$
$$t > 0 \quad x \to \infty \quad T = T_\infty$$

It is desired to use an aluminium plate to provide an isothermal wall for periods not exceeding 10 seconds in an enclosure subjected to transients. What thickness of the plate is required if the temperature in this plate is not to be less than 95 % of the surface temperature. Assume that the initial temperature is 303 K and the surface temperature 750 K.

Answer: For Aluminium, $\rho = 2700 \text{ kg/m}^3$; $Cp = 938 \text{ J/kg.K}$; $k = 228 \text{ W/m.K}$; $T = 0.95T_S = 712.5 \text{ K}$

$$\alpha = \frac{k}{\rho Cp} = \frac{228}{2700 \times 938}, \frac{W}{mK} \frac{m^3}{kg} \frac{kgK}{J} = 9.00 \times 10^{-5}, \frac{m^2}{s}$$

Substituting the given values into the given equation

$$\frac{712.5 - 303}{750 - 303} = 1 - erf\left(\frac{x}{2\sqrt{9.00 \times 10^{-5} \times 10}}\right) = 0.9161$$

That is

$$erf\left(\frac{x}{0.06}\right) = 1 - 0.9161 = 0.0839$$

From Error Function tables

$$\frac{x}{0.06} = 0.0745$$

which gives x = 0.0045m = 4.5mm Ans.

Example 7.23: The solution to the transient one dimensional heat conduction equation for an isotropic solid

$$\frac{\partial T}{\partial t} = \alpha \frac{\partial^2 T}{\partial x^2}$$

for the case of a fluid flowing past the surface of a wall with a film heat transfer coefficient, h, at constant temperature, T_f, is given by

$$\frac{T_f - T}{T_f - T_o} = erf\left(\frac{x}{2\sqrt{\alpha t}}\right) + exp\left(\frac{hx}{k} + \frac{h^2 \alpha t}{k^2}\right)\left[1 - erf\left(\frac{h\sqrt{\alpha t}}{k} + \frac{x}{2\sqrt{\alpha t}}\right)\right]$$

where T = temperature of the solid at time t and at a distance x, T_0 = initial temperature; α, k = thermal diffusivity and conductivity, respectively.

Determine the temperature, after 3 minutes, in a plane, 6 mm deep from the surface of an infinite aluminium plate, initially at 30 C, past which fluid, at 95 C, is suddenly pumped. Take h = 800 W/m²K

Answer: The properties of aluminium plate are obtained from Tables as follows: ρ = 2700 kg/m³; Cp = 938 J/kg.K; k = 228 W/m.K

Hence

$$\alpha = \frac{k}{\rho Cp} = \frac{228}{2700 \ x \ 938}, \frac{W}{mK} \frac{m^3}{kg} \frac{kgK}{J} = 9.00 \ x \ 10^{-5}, \frac{m^2}{s} \quad (1)$$

Also $t = 3 \ x \ 60 = 180 \ s; x = 0.006 \ m; \ T_o = 30 + 273 = 303 \ K;$

$$T_f = 95 + 273 = 368 \ K \quad (2)$$

$$\frac{x}{2\sqrt{\alpha t}} = \frac{0.006}{2\sqrt{9.00 \ x \ 10^{-5} \ x \ 180}} = 0.02357 \quad (3)$$

From the tables of the Error Function

$$erf\left(\frac{x}{2\sqrt{\alpha t}}\right) = erf(0.02357) = 0.021 \quad (4)$$

$$\frac{hx}{k} + \frac{h^2\alpha t}{k^2} = \frac{800 \; x \; 0.006}{228}, \frac{W}{m^2 K} \frac{mK}{W} m$$

$$+ \frac{(800)^2 \; x \; 9.00 \; x \; 10^{-5} \; x \; 180}{(228)^2}, \frac{W^2}{m^4 K^2} \frac{m^2}{s} \frac{m^2 K^2}{W^2} s$$

$$= 0.0211 + 0.1995 = 0.2206 \tag{5}$$

This gives

$$exp\left(\frac{hx}{k} + \frac{h^2\alpha t}{k^2}\right) = exp(0.2206) = 1.2468 \tag{6}$$

Also

$$\frac{h\sqrt{\alpha t}}{k} + \frac{x}{2\sqrt{\alpha t}} = \frac{800\sqrt{9.00 \; x \; 10^{-5} \; x \; 180}}{228} + \frac{0.006}{2\sqrt{9.00 \; x \; 10^{-5} \; x \; 180}}$$

$$= 0.4466 + 0.0236 = 0.4702 \tag{7}$$

From the tables of the Error Function

$$erf\left(\frac{h\sqrt{\alpha t}}{k} + \frac{x}{2\sqrt{\alpha t}}\right) = erf(0.4702) = 0.444 \tag{8}$$

Substituting (2), (4), (6) and (8) into the given equation

$$\frac{T_f - T}{T_f - T_o} = \frac{368 - T}{368 - 303} = 0.021 + 1.247(1 - 0.444) = 0.7143$$

we get that

$$368 - T = 46.4 \quad or \quad T = 321.6 \; K \quad or \quad 48.6 \; C. \quad Ans$$

Example 7.24: The implicit form of the unsteady state heat conduction equation, with internal heat generation, in an isotropic solid is given, using the heat balance method, by

$$T_i^{n+1} = \frac{1}{\left[1 + \sum_{j=1}^{N}\frac{K_{ij}}{C_i}\right]^{n+1}}\left[T_i^n + \left(\frac{q_iV_i}{C_i}\right)^{n+1} + \sum_{j=1}^{N}\left(\frac{K_{ij}}{C_i}T_j\right)^{n+1}\right] \quad (1)$$

where T_i^n and T_i^{n+1} are the temperatures at node i and at times n and $n+1$ respectively; V_i, C_i, q_i are the volume, modified heat capacity and the heat flux, respectively, in node i.

$$C_i = \frac{\rho_i V_i C p_i}{\Delta t} \quad (1a)$$

where ρ_i is the density at node i; Cp_i is the heat capacity at node i, and

$$K_{ij} = \frac{1}{R_{ij} + R_{ji}} = conductance\ between\ nodes\ i\ and\ j \quad (1b)$$

where R_{ij} and R_{ji} are the resistances to heat transfer from nodes i to j and j to i, respectively.

$$R_{ij} = \frac{\delta_{ij}}{k_i A_{ij}} \quad (1c)$$

δ_{ij} is the half node thickness between nodes i and j as shown below, k_i is the thermal conductivity at node i and A_{ij} is the area, perpendicular to the heat flux, between nodes i and j.

A two dimensional mild steel plate, 0.6m square, of infinite length, and, initially, at 303 K has three of its sides insulated while the fourth side exchanges heat with another surface at constant temperature. If this plate has an internal heater, generating heat at the rate of 20 kW/m³, determine the temperature distribution within the plate in the first 100 seconds. Take this constant temperature to be 380 K and the heat transfer coefficient at the isothermal surface to be 200 W/m²K.

Answer: The physical situation is as shown below

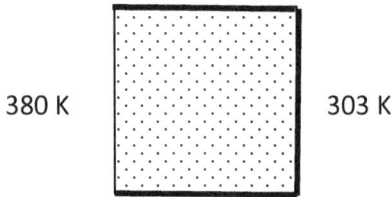

Let us divide the plate into four nodes to minimise the tedium of hand calculation. (In actual situations, using computer algorithms, more nodes may need to be selected for greater accuracy). The four nodes on the plate, insulated on three sides, are then as shown below (1 - 4)

Nodal Arrangement on Plate Insulated on Three Sides

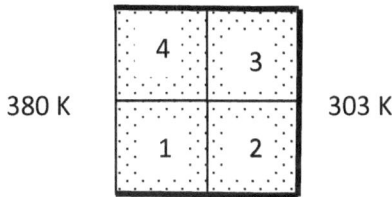

For the purpose of the calculations, however, the four actual nodes selected on the plate will have to be surrounded by imaginary nodes, the reasons for which will be apparent soon enough. These new and surrounding nodal arrangements are illustrated as shown below.

If we let $\Delta x = \Delta y = 0.3$m then $\delta_{ij} = 0.15$m. The relevant material properties are obtained from standard tables as

	Mild Steel	**Asbestos**	**Water**
ρ	7820 kg/m^3	577 kg/m^3	996 kg/m^3
Cp	473 J/kgK	1050 J/kgK	4183 J/kgK
k	42.9 W/mK	0.157 W/mK	0.611 W/mK

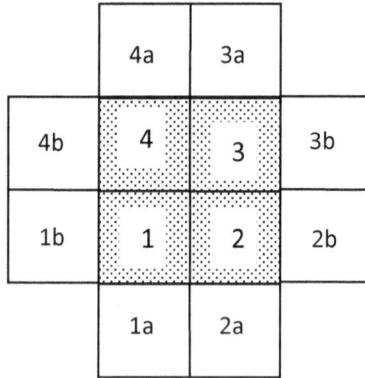

For the mild steel plate

$$V_i = 0.3 \; x \; 0.3 \; x \; 1 = 0.09 \; m^3/m \; of \; depth$$

Taking Δt = 100s

$$C_i = \frac{\rho_i V_i C p_i}{\Delta t} = \frac{7820 \; x \; 0.09 \; x \; 473}{100} \frac{kg}{m^3} \frac{J}{kgK} m^3 \frac{1}{s}$$
$$= 3328.97 \; W/K$$

For i = 1

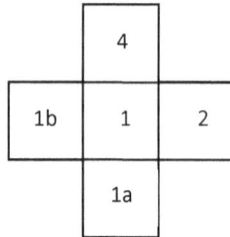

$$A_{1j} = 0.3 \; x \; 1 = 0.3 \; m^2$$

$$R_{1j} = \frac{\delta_{1j}}{k_1 A_{1j}} = \frac{0.15}{42.9 \; x \; 0.3}, m \frac{mK}{W} \frac{1}{m^2} = 0.01166 \; K/W$$

except for

$$R_{1b1} = \frac{1}{h_1 A_{12}} = \frac{1}{200 \times 0.3}, \frac{m^2 K}{W} \frac{1}{m^2} = 0.01667 \ K/W$$

and

$$R_{1a1} = \frac{1}{h_{1a} A_{11}} = \frac{1}{0 \times 0.3}, \frac{m^2 K}{W} \frac{1}{m^2} = \infty \ K/W$$

Thus, for i = 1, all j

$$R_{11a} = 0.01166 \qquad R_{12} = 0.01166 \qquad R_{14} = 0.01166$$

$$R_{1a1} = \infty \qquad R_{21} = 0.01166 \qquad R_{41} = 0.01166$$

$$R_{11b} = 0.01166 \qquad R_{1b1} = 0.01667 \qquad R_{11} = 0.01166$$

Also

$$K_{11} = \frac{1}{R_{11} + R_{11}} = \frac{1}{0.01166 + 0.01166} = 42.88 \ W/K$$

$$K_{12} = \frac{1}{R_{12} + R_{21}} = \frac{1}{0.01166 + 0.01166} = 42.88 \ W/K = K_{21}$$

$$K_{14} = \frac{1}{R_{14} + R_{41}} = \frac{1}{0.01166 + 0.01166} = 42.88 \ W/K = K_{41}$$

$$K_{11a} = \frac{1}{R_{11a} + R_{1a1}} = \frac{1}{0.01166 + \infty} = 0 \ W/K = K_{1a1}$$

$$K_{11b} = \frac{1}{R_{11b} + R_{1b1}} = \frac{1}{0.01166 + 0.01667} = 35.298 \ W/K = K_{41}$$

Then

$$\sum_{j=1}^{4} \frac{K_{1j}}{C_1} = \frac{1}{C_1}(K_{11} + K_{12} + K_{14} + K_{11b} + K_{11a})$$

$$= \frac{1}{3328.97}(42.88 + 42.88 + 42.88 + 35.298 + 0), \frac{W\ K}{K\ W}$$
$$= 0.0492$$

$$\sum_{j=1}^{4} \frac{K_{1j}}{C_1} T_j = \frac{1}{C_1}(K_{11}T_1 + K_{12}T_2 + K_{14}T_4 + K_{11b}T_{1b})$$

$$= (0.0129T_1 + 0.0129T_2 + 0.0129T_4 + 0.0106T_{1b}), \frac{W\ K}{K\ W} K \quad (2)$$

$$\frac{q_1 V_1}{C_1} = \frac{20{,}000\ x\ 0.09}{3328.97}, \frac{W}{m^2} \frac{m^3}{m} \frac{K}{W} = 0.5407\ K \quad (3)$$

$$\frac{1}{1 + \sum_{j=1}^{4}\left(\frac{K_{1j}}{C_1}\right)} = \frac{1}{1.0492} = 0.9531 \quad (4)$$

Substituting (2), (3) and (4) in (1)

$$T_1^{n+1} = 0.9531\big[T_1^n + (0.5407)^{n+1} + 0.0129T_1^{n+1}$$
$$+ 0.0129T_2^{n+1} + 0.0129T_4^{n+1}$$
$$+ 0.0106T_{1b}^{n+1}\big] \quad (5)$$

At $n = 0$

$$T_1^1 = 0.9531\big[T_1^0 + (0.5407)^1 + 0.0129T_1^1 + 0.0129T_2^1$$
$$+ 0.0129T_4^1 + 0.0106T_{1b}^1\big]$$

$$= 0.9531T_1^0 + 0.5153 + 0.0123T_1^1 + 0.0123T_2^1 + 0.0123T_4^1$$
$$+ 0.0101T_{1b}^1$$

which can be rearranged as

$$0.9871T_1^1 - 0.0123T_2^1 - 0.0123T_4^1 - 0.0101T_{1b}^1$$
$$= 0.9531T_1^0 + 0.5153 \quad (6)$$

For i = 2

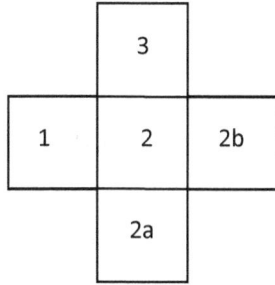

$$A_{2j} = 0.3 \; x \; 1 = 0.3 \; m^2$$

$$R_{2j} = \frac{\delta_{2j}}{k_2 A_{2j}} = \frac{0.15}{42.9 \; x \; 0.3}, m \frac{mK}{W} \frac{1}{m^2} = 0.01166 \; K/W$$

except for

$$R_{2a2} = \frac{1}{h_{2a} A_{2a2}} = \frac{1}{0 \; x \; 0.3}, \frac{m^2 K}{W} \frac{1}{m^2} = \infty \; K/W$$

and

$$R_{2b2} = \frac{1}{h_{2b} A_{2b2}} = \frac{1}{0 \; x \; 0.3}, \frac{m^2 K}{W} \frac{1}{m^2} = \infty \; K/W$$

Thus, for i = 2, all j

$R_{21} = 0.01166$	$R_{12} = 0.01166$	$R_{23} = 0.01166$
$R_{32} = 0.01166$	$R_{2a2} = \infty$	$R_{22a} = 0.01166$
$R_{2b2} = \infty$	$R_{22b} = 0.01166$	$R_{22} = 0.01166$

Also

$$K_{11} = \frac{1}{R_{11} + R_{11}} = \frac{1}{0.01166 + 0.01166} = 42.88 \; W/K = K_{22}$$

$$K_{12} = \frac{1}{R_{12} + R_{21}} = \frac{1}{0.01166 + 0.01166} = 42.88 \; W/K = K_{21}$$

$$K_{23} = \frac{1}{R_{23} + R_{32}} = \frac{1}{0.01166 + 0.01166} = 42.88 \ W/K$$

$$K_{22a} = \frac{1}{R_{22a} + R_{2a2}} = \frac{1}{0.01166 + \infty} = 0 \ W/K$$

$$K_{22b} = \frac{1}{R_{22b} + R_{2b2}} = \frac{1}{0.01166 + \infty} = 0 \ W/K$$

Then

$$\sum_{j=1}^{4} \frac{K_{2j}}{C_2} = \frac{1}{C_2}(K_{21} + K_{22} + K_{22a} + K_{22b} + K_{23})$$

$$= \frac{1}{3328.97}(42.88 + 42.88 + 0 + 0 + 42.88), \frac{W}{K}\frac{K}{W} = 0.0386$$

$$\sum_{j=1}^{4} \frac{K_{2j}}{C_2}T_j = \frac{1}{C_2}(K_{21}T_1 + K_{22}T_2 + K_{23}T_3)$$

$$= (0.0129T_1 + 0.0129T_2 + 0.0129T_3), \frac{W}{K}\frac{K}{W}K \quad (7)$$

$$\frac{q_2 V_2}{C_2} = \frac{20{,}000 \ x \ 0.09}{3328.97}, \frac{W}{m^2}\frac{m^3}{m}\frac{K}{W} = 0.5407 \ K \quad (8)$$

$$\frac{1}{1 + \sum_{j=1}^{4}\left(\frac{K_{2j}}{C_2}\right)} = \frac{1}{1.0386} = 0.9628 \quad (9)$$

Substituting (7), (8) and (9) in (1)

$$T_2^{n+1} = 0.9628\big[T_2^{n} + (0.5407)^{n+1} + 0.0129T_1^{n+1}$$
$$+ 0.0129T_2^{n+1} + 0.0129T_3^{n+1}$$
$$+ 0\big] \quad (10)$$

At n = 0

$$T_2{}^1 = 0.9628\big[T_2{}^0 + (0.5407)^1 + 0.0129T_1{}^1 + 0.0129T_2{}^1 + 0.0129T_3{}^1 + 0\big]$$

$$= 0.9628T_2{}^0 + 0.5203 + 0.0124T_1{}^1 + 0.0124T_2{}^1 + 0.0124T_3{}^1$$

which can be rearranged as

$$-0.0124T_1{}^1 + 0.9876T_2{}^1 - 0.0124T_3{}^1 = 0.9628T_2{}^0$$
$$+ 0.5203 \qquad (11)$$

For i = 3

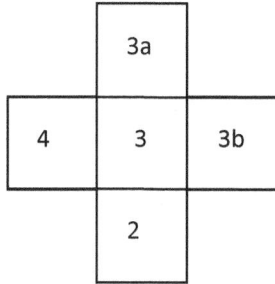

$$A_{3j} = 0.3 \times 1 = 0.3 \; m^2$$

$$R_{3j} = \frac{\delta_{3j}}{k_3 A_{3j}} = \frac{0.15}{42.9 \times 0.3}, m\frac{mK}{W}\frac{1}{m^2} = 0.01166 \; K/W$$

except for

$$R_{3a3} = \frac{1}{h_{3a}A_{3a3}} = \frac{1}{0 \times 0.3}, \frac{m^2 K}{W}\frac{1}{m^2} = \infty \; K/W$$

and

$$R_{3b3} = \frac{1}{h_{2b}A_{2b2}} = \frac{1}{0 \times 0.3}, \frac{m^2 K}{W}\frac{1}{m^2} = \infty \; K/W$$

Thus, for i = 3, all j

$$R_{32} = 0.01166 \qquad R_{23} = 0.01166 \qquad R_{33b} = 0.01166$$

$$R_{33a} = 0.01166 \qquad R_{3a3} = \infty \qquad R_{34} = 0.01166$$

$$R_{3b3} = \infty \qquad R_{43} = 0.01166 \qquad R_{33} = 0.01166$$

Also

$$K_{33} = \frac{1}{R_{33} + R_{33}} = \frac{1}{0.01166 + 0.01166} = 42.88 \ W/K = K_{22}$$

$$K_{32} = \frac{1}{R_{32} + R_{23}} = \frac{1}{0.01166 + 0.01166} = 42.88 \ W/K = K_{21}$$

$$K_{34} = \frac{1}{R_{34} + R_{43}} = \frac{1}{0.01166 + 0.01166} = 42.88 \ W/K$$

$$K_{33a} = \frac{1}{R_{33a} + R_{3a3}} = \frac{1}{0.01166 + \infty} = 0 \ W/K$$

$$K_{33b} = \frac{1}{R_{33b} + R_{3b3}} = \frac{1}{0.01166 + \infty} = 0 \ W/K$$

Then

$$\sum_{j=1}^{4} \frac{K_{3j}}{C_3} = \frac{1}{C_3}(K_{32} + K_{33} + K_{34} + K_{33a} + K_{33b})$$

$$= \frac{1}{3328.97}(42.88 + 42.88 + 42.88 + 0 + 0), \frac{W}{K}\frac{K}{W} = 0.0386$$

$$\sum_{j=1}^{4} \frac{K_{3j}}{C_3} T_j = \frac{1}{C_3}(K_{32}T_2 + K_{33}T_3 + K_{34}T_4)$$

$$= (0.0129T_2 + 0.0129T_3 + 0.0129T_4), \frac{W}{K}\frac{K}{W}K \quad (12)$$

$$\frac{q_3V_3}{C_3} = \frac{20,000 \ x \ 0.09}{3328.97}, \frac{W}{m^2}\frac{m^3}{m}\frac{K}{W} = 0.5407 \ K \quad (13)$$

$$\frac{1}{1 + \sum_{j=1}^{4}\left(\frac{K_{3j}}{C_3}\right)} = \frac{1}{1.0386} = 0.9628 \qquad (14)$$

Substituting (12), (13) and (14) in (1)

$$T_3{}^{n+1} = 0.9628\left[T_3{}^n + (0.5407)^{n+1} + 0.0129T_2{}^{n+1}\right. $$
$$\left. + 0.0129T_3{}^{n+1} + 0.0129T_4{}^{n+1} + 0\right] \qquad (15)$$

At n = 0

$$T_3{}^1 = 0.9628\left[T_3{}^0 + (0.5407)^1 + 0.0129T_2{}^1 + 0.0129T_3{}^1\right.$$
$$\left. + 0.0129T_4{}^1\right]$$

$$= 0.9628T_3{}^0 + 0.5203 + 0.0124T_2{}^1 + 0.0124T_3{}^1 + 0.0124T_4{}^1$$

which can be rearranged as

$$-0.0124T_2{}^1 + 0.9876T_3{}^1 - 0.0124T_4{}^1 = 0.9628T_3{}^0$$
$$+ 0.5203 \qquad (16)$$

For i = 4

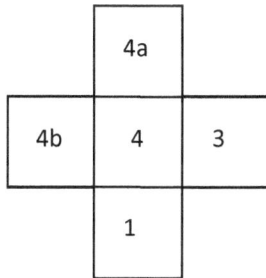

$$A_{4j} = 0.3 \; x \; 1 = 0.3 \; m^2$$

$$R_{4j} = \frac{\delta_{4j}}{k_4 A_{4j}} = \frac{0.15}{42.9 \; x \; 0.3}, m\frac{mK}{W}\frac{1}{m^2} = 0.01166 \; K/W$$

except for

$$R_{4a4} = \frac{1}{h_{4a}A_{4a4}} = \frac{1}{0 \times 0.3}, \frac{m^2 K}{W}\frac{1}{m^2} = \infty \ K/W$$

and

$$R_{4b4} = \frac{1}{h_{4b}A_{4b4}} = \frac{1}{200 \times 0.3}, \frac{m^2 K}{W}\frac{1}{m^2} = 0.01667 \ K/W$$

Thus, for i = 4, all j

$R_{41} = 0.01166$	$R_{14} = 0.01166$	$R_{43} = 0.01166$
$R_{34} = 0.01166$	$R_{4a4} = \infty$	$R_{44a} = 0.01166$
$R_{44b} = 0.01166$	$R_{4b4} = 0.01667$	$R_{44} = 0.01166$

Also

$$K_{41} = \frac{1}{R_{41} + R_{14}} = \frac{1}{0.01166 + 0.01166} = 42.88 \ W/K = K_{14}$$

$$K_{43} = \frac{1}{R_{43} + R_{34}} = \frac{1}{0.01166 + 0.01166} = 42.88 \ W/K = K_{34}$$

$$K_{44} = \frac{1}{R_{44} + R_{44}} = \frac{1}{0.01166 + 0.01166} = 42.88 \ W/K$$

$$K_{44a} = \frac{1}{R_{44a} + R_{4a4}} = \frac{1}{0.01166 + \infty} = 0 \ W/K$$

$$K_{44b} = \frac{1}{R_{44b} + R_{4b4}} = \frac{1}{0.01166 + 0.01667} = 35.298 \ W/K$$

Then

$$\sum_{j=1}^{4} \frac{K_{4j}}{C_4} = \frac{1}{C_4}(K_{41} + K_{43} + K_{44} + K_{44a} + K_{44b})$$

$$= \frac{1}{3328.97}(42.88 + 42.88 + 42.88 + 0 + 35.298), \frac{W}{K}\frac{K}{W}$$
$$= 0.0492$$

$$\sum_{j=1}^{4} \frac{K_{4j}}{C_4} T_j = \frac{1}{C_4}(K_{41}T_1 + K_{43}T_3 + K_{44}T_4 + K_{44a}T_{4a} + K_{44b}T_{4b})$$

$$= (0.0129T_1 + 0.0129T_3 + 0.0129T_4 + 0.0106T_{4b}), \frac{W}{K}\frac{K}{W}K \quad (17)$$

$$\frac{q_4 V_4}{C_4} = \frac{20,000 \times 0.09}{3328.97}, \frac{W}{m^2}\frac{m^3}{m}\frac{K}{W} = 0.5407 \, K \quad (18)$$

$$\frac{1}{1 + \sum_{j=1}^{4}\left(\frac{K_{4j}}{C_4}\right)} = \frac{1}{1.0492} = 0.9531 \quad (19)$$

Substituting (17), (18) and (19) in (1)

$$T_4^{n+1} = 0.9531\big[T_4^{n} + (0.5407)^{n+1} + 0.0129T_1^{n+1}$$
$$+ 0.0129T_3^{n+1} + 0.0129T_4^{n+1}$$
$$+ 0.0106T_{4b}^{n+1}\big] \quad (20)$$

At n = 0

$$T_4^{1} = 0.9531\big[T_4^{0} + (0.5407)^{1} + 0.0129T_1^{1} + 0.0129T_3^{1}$$
$$+ 0.0129T_4^{1} + 0.0106T_{4b}^{1}\big]$$

$$= 0.9531T_4^{0} + 0.5153 + 0.0123T_1^{1} + 0.0123T_3^{1} + 0.0123T_4^{1}$$
$$+ 0.0101T_{4b}^{1}$$

which can be rearranged as

$$-0.0123T_1^{1} - 0.0123T_3^{1}$$
$$+ 0.9877T_4^{1} = 0.0101T_{4b}^{1} + 0.9531T_3^{0}$$
$$+ 0.5153 \quad (21)$$

Listing equations (6), (11), (16) and (21) together, we get

$$0.9871T_1{}^1 - 0.0123T_2{}^1 - 0.0123T_4{}^1 - 0.0101T_{1b}{}^1$$
$$= 0.9531T_1{}^0 + 0.5153 \qquad (6)$$

$$-0.0124T_1{}^1 + 0.9876T_2{}^1 - 0.0124T_3{}^1 = 0.9628T_2{}^0$$
$$+ 0.5203 \qquad (11)$$

$$-0.0124T_2{}^1 + 0.9876T_3{}^1 - 0.0124T_4{}^1 = 0.9628T_3{}^0$$
$$+ 0.5203 \qquad (16)$$

$$-0.0123T_1{}^1 - 0.0123T_3{}^1$$
$$+ 0.9877T_4{}^1 = 0.0101T_{4b}{}^1 + 0.9531T_3{}^0$$
$$+ 0.5153 \qquad (21)$$

Since $T_1^o = 303 = T_2^o = T_3^o = T_4^o$ and $T_{1b} = T_{4b} = 380$, equations (6), (11), (16) and (21) simplify to

$$0.9871T_1{}^1 - 0.0123T_2{}^1 - 0.0123T_4{}^1$$

$$= 0.0101 \; x \; 380 + 0.9531 \; x \; 303 + 0.5153 = 293.14 \quad (22)$$

$$-0.0124T_1{}^1 + 0.9876T_2{}^1 - 0.0124T_3{}^1 = 0.9628 \; x \; 303 + 0.5203$$
$$= 292.25 \qquad (23)$$

$$-0.0124T_2{}^1 + 0.9876T_3{}^1 - 0.0124T_4{}^1 = 0.9628 \; x \; 303 + 0.5203$$
$$= 292.25 \qquad (24)$$

$$-0.0123T_1{}^1 - 0.0123T_3{}^1 + 0.9877T_4{}^1$$
$$= 0.0101 \; x \; 380 + 0.9531 \; x \; 303 + 0.5153$$
$$= 293.14 \qquad (25)$$

Equations (22), (23), (24) and (25) result in the matrix

$$\begin{pmatrix} 0.9871 & -0.0123 & 0 & -0.0123 \\ -0.0124 & 0.9876 & -0.0124 & 0 \\ 0 & -0.0124 & 0.9876 & -0.0124 \\ -0.0123 & 0 & -0.0123 & 0.9877 \end{pmatrix} \begin{pmatrix} T_1^{\ 1} \\ T_2^{\ 1} \\ T_3^{\ 1} \\ T_4^{\ 1} \end{pmatrix} = \begin{pmatrix} 293.14 \\ 292.25 \\ 292.25 \\ 293.14 \end{pmatrix} \quad (26)$$

The MATLAB^R solution of equation (26) is

$$T_1^{\ 1} = 304.36 \, K$$
$$T_2^{\ 1} = 303.55K$$
$$T_3^{\ 1} = 303.55 \, K$$
$$T_4^{\ 1} = 304.36 \, K$$

Note that for subsequent time intervals such as the next 100 seconds, equations (6), (11), (16) and (21) are used with $T_1^{\ 1}$, $T_2^{\ 1}, T_3^{\ 1}, T_4^{\ 1}$ replacing $T_1^{\ o}, T_2^{\ o}, T_3^{\ o}, T_4^{\ o}$ and so on.

References For Chapter Seven

1   Carslaw, H.S and Jaeger, J.C., *Conduction of Heat in Solids*, Clarendon Press, Oxford, UK, 1959

2   Welty J.R: *Engineering Heat Transfer*, John Wiley and Sons New York, USA. 1978

3   Zill D. G. and Cullen M. R., *Differential Equations with Boundary Value Problems*; Brooks/Cole Publishing Co., California, USA, 1997

4   MATLAB ; Version 7,0,4 365(R14) Service Pack 2, January 29, 2005; © 1994-2005 The MathWorks, Inc.

CHAPTER EIGHT
NON-STEADY STATE, MULTI-DIMENSIONAL, HEAT CONDUCTION

Example 8.01: What is the relevance of multidimensional systems in heat conduction?

Answer: The relevance of multidimensional systems in heat conduction is that reality, as we know and experience it, is multidimensional. The first seven chapters of this book have highlighted the human approach to dealing with this reality, at least, in heat conduction. This approach is that of segmentation and compartmentalisation of real multidimensional problems into one, two or more dimensions depending on what the available methods of analysis can handle.

Einstein's analysis of relativity was the first to show that reality is more than three dimensions. Before that, however, mathematical analyses have always shown, in matrix algebra, the existence of n- dimensions, even if in abstract or hyper space. Economic forecasts and analysis, understanding human behaviour, weather forecasts, for example, are areas where multidimensional analysis is yielding fruit.

In heat conduction, however, we have enough problems, as it is, dealing with one and two space dimensions plus the time dimension. Nevertheless, procedures have been developed for three space dimensional heat conduction problems in which the fourth dimension, time, is, also, involved. Only one of these procedures is illustrated below.

Example 8.02: Outline the non-steady state, energy equations useful for solving problems in multi-dimensional non-steady state conduction.

213

Answer: Two situations are important, namely, when there is no internal generation of heat and when there is internal generation of heat. Commercial interest is wary of multi-dimensional analysis because of cost and complexity unless there is no other way of dealing with the problem. In such cases, a vector mathematical approach is, usually, preferred. Only scalar procedures are illustrated here, however, in keeping with the purpose of this book to bring beginners up to speed. Only systems without internal generation of heat will, also for the same reasons, be illustrated.

<u>Case 1: No Internal Generation of Heat, Qv = 0</u>

In two space dimensions, x and y, equation (2.07) becomes

$$\frac{\partial T}{\partial t} = \frac{k}{\rho Cp}\left(\frac{\partial^2 T}{\partial x^2} + \frac{\partial^2 T}{\partial y^2}\right) \tag{8.01}$$

In three space dimensions, it is

$$\frac{\partial T}{\partial t} = \frac{k}{\rho Cp}\left(\frac{\partial^2 T}{\partial x^2} + \frac{\partial^2 T}{\partial y^2} + \frac{\partial^2 T}{\partial z^2}\right) \tag{8.02}$$

<u>Case 2: With Internal Generation of Heat</u>

In two space dimensions, equation (2.06) becomes

$$\rho Cp\,\frac{\partial T}{\partial t} = Q_V + k\left(\frac{\partial^2 T}{\partial x^2} + \frac{\partial^2 T}{\partial y^2}\right) \tag{8.03}$$

while in three space dimensions, it is

$$\rho Cp\,\frac{\partial T}{\partial t} = Q_V + k\left(\frac{\partial^2 T}{\partial x^2} + \frac{\partial^2 T}{\partial y^2} + \frac{\partial^2 T}{\partial z^2}\right) \tag{8.04}$$

Example 8.03: Outline the methods for solving the above equations of multi-dimensional, unsteady state heat conduction.

Answer: Case 1: No internal generation of heat, $Q_V = 0$

From equation (8.01)

$$\frac{\partial T}{\partial t} = \frac{k}{\rho C p}\left(\frac{\partial^2 T}{\partial x^2} + \frac{\partial^2 T}{\partial y^2}\right) \qquad (8.01)$$

According to Welty (1978), this can be reduced to

$$\frac{\partial Y}{\partial t} = \alpha\left(\frac{\partial^2 Y}{\partial x^2} + \frac{\partial^2 Y}{\partial y^2}\right) \qquad (8.05)$$

where

$$\alpha = \frac{k}{\rho C p}$$

and

$$Y = Y(x, y, t) = \frac{T(x,y,t) - T_\infty}{T_0 - T_\infty} \qquad (8.05a)$$

with initial conditions given by

$$t = 0, \qquad Y(x, y, 0) = \frac{T_0 - T_\infty}{T_0 - T_\infty} = 1 \qquad (8.05b)$$

and boundary conditions at t > 0 given by

$x = 0$	$\dfrac{\partial Y(0, y, t)}{\partial x} = 0$	(8.05c)
$y = 0$	$\dfrac{\partial Y(x, 0, t)}{\partial y} = 0$	(8.05d)
$x = L$	$Y(L, y, t) + \dfrac{k}{h}\dfrac{\partial Y(L, y, t)}{\partial x} = 0$	(8.05e)
$y = L$	$(x, L, t) + \dfrac{k}{h}\dfrac{\partial Y(x, L, t)}{\partial y} = 0$	(8.05f)

By the separation of variables

$$Y(x, y, t) = X(x, t).Y(y, t) \qquad (8.06)$$

equation (8.05) becomes

$$X\frac{\partial Y}{\partial t} + Y\frac{\partial X}{\partial t} = \alpha \left(Y\frac{\partial^2 X}{\partial x^2} + X\frac{\partial^2 Y}{\partial y^2} \right) \qquad (8.07)$$

Dividing equation (8.07) by XY, we get

$$\frac{1}{X}\left[\frac{\partial X}{\partial t} - \alpha\frac{\partial^2 X}{\partial x^2} \right] = -\frac{1}{Y}\left[\frac{\partial Y}{\partial t} - \alpha\frac{\partial^2 Y}{\partial y^2} \right] = 0 \qquad (8.08)$$

Hence it is seen that solving equation (8.08) is equivalent to the solution of two one dimensional problems with

$$X = \frac{T(x, t) - T_\infty}{T_o - T_\infty} \qquad (8.08a)$$

$$Y = \frac{T(y, t) - T_\infty}{T_o - T_\infty} \qquad (8.08b)$$

This means that the solutions for two, or even three, dimensional problems are the products of the one dimensional solution. For two and three dimensional objects that are finite such as a rectangular bar with insulated ends

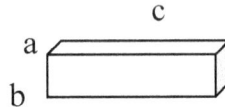

In two dimensions, $\quad Y_{bar} = Y_a.Y_b \qquad (8.08c)$

The Heissler chart (see Appendix 1) is used to evaluate the Ys as functions of x and y. For a rectangular parallelepiped, for example, Y is given, for heat conduction in all directions, as

216

$$Y_{bar} = Y_a \cdot Y_b \cdot Y_c \qquad (8.08d)$$

For a cylinder (including both ends)

$$Y_c = Y_{cylinder} \cdot Y_a \qquad (8.08e)$$

and Y_a is evaluated from the chart assuming that each end of the cylinder is like a flat circular plate of radius, $x_i = a$

For heat transport from only one face of a flat plate, a value of x, which is twice its true value, is used. This is equivalent to using one face of the plate as the plane of symmetry. Various cases are treated in Carslaw & Jaeger (1959).

Example 8.04: Outline other methods for solving the above equations of multi-dimensional, unsteady state heat conduction, especially if the object is not of regular shape.

Answer: Food technologists express the equation which describes the lumped parameter system (negligible internal resistance)

$$\frac{T - T_\infty}{T_o - T_\infty} = e^{-\frac{h_s A t}{\rho V C p}}$$

as

$$log(T - T_\infty) = -\frac{t}{f_h} + \log j \cdot (T_o - T_\infty) \qquad (8.09)$$

where

$$f_h = 2.303 \frac{\rho V C p}{h_c A t}$$

$$= time\ factor\ rate\ of\ heating\ or\ cooling \qquad (8.09a)$$

and

$$j = \frac{T_a - T_\infty}{T_o - T_\infty} = lag\ factor \qquad (8.09b)$$

h_C = heat transfer coefficient for cooling
T_a = intercept on the log plot of the cooling curve

Only centreline temperatures are used. f_h and j are, then, determined for various shapes as follows:

For a finite parallelepiped (p)

$$\frac{1}{f_p} = \frac{1}{f_{length}} + \frac{1}{f_{width}} + \frac{1}{f_{height}} \qquad (8.10)$$

$$j_p = j_{length} \cdot j_{width} \cdot j_{height} \qquad (8.11)$$

For a finite cylinder (f_c)

$$\frac{1}{f_{fc}} = \frac{1}{f_{infinite\ cylinder}} + \frac{1}{f_{height}} \qquad (8.12)$$

$$j_{fc} = j_{infinite\ cylinder} \cdot j_{height} \qquad (8.13)$$

Irregularly Shaped Objects and Ellipsoids

Some objects, in practical use, do not fit neatly into the shapes of rectangular parallelepipeds, cylinders or perfectly rounded spheres. Sometimes it is possible to use shape factors, the most common are those which define the shape factors with respect to the rectangular parallelepiped, cylinder or sphere which has the same surface area or volume as the actual object. In such cases, the equations, shown above, can be used.

Where the object can be approximated to an ellipsoid, an index, G, can be defined. Thus, for an ellipsoid shown below, when $0.25 < G < 1$

$$G = \frac{1}{4} + \frac{3}{8A^2} + \frac{3}{8B^2} \qquad (8.14)$$

where $A = a/z$, and $B = b/z$. a and b are the half length and half width, respectively of the ellipsoid while z is a characteristic dimension or half the minimum thickness.

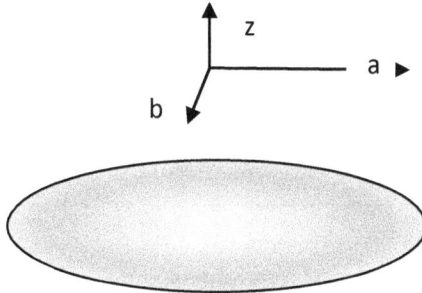

For example, $G = 0.25$ for an infinite slab, 0.585 for a cylinder and 1.0 for a sphere.

The temperature distribution is determined from the Heissler charts with the mass average temperature located at

$$L_{mass\ average} = G^{0.14} - 0.25 \qquad (8.15)$$

and

$$L_{mass\ average} = \frac{desired\ length\ along\ characteristic\ length}{characteristic\ length\ (z)} \qquad (8.16)$$

Example 8.05: A carrot, whose shape may be approximated by a cylinder 0.2m long and 19mm diameter, is initially at room temperature, 295K, and then dropped into boiling water at atmospheric pressure. Properties of the system and the carrot may be taken as follows: h = 2000 W/m²K; k = 0.48 W/mK; Cp = 4.0 kJ/kg.K; ρ = 1025 kg/m³. How long must the carrot cook if the requirement is that the minimum temperature reached be 365K?

Answer: It is necessary to determine, first, the Blot's number. For a finite cylinder, where all surfaces are involved

$$Bi = \frac{hV}{kA} = \frac{h}{k} \left[\frac{\pi \left(\frac{D^2}{4}\right) L}{2\pi \frac{D^2}{4} + \pi DL} \right] = \frac{h}{k} \left(\frac{L \cdot D}{2D + 4L} \right) \qquad (1)$$

That is

$$Bi = \frac{2000}{0.48} \left(\frac{0.2 \times 0.019}{2 \times 0.019 + 4 \times 0.2} \right), \frac{W}{m^2 K} \frac{mK}{W} \frac{m^2}{m} = 18.89 \qquad (2)$$

That is $0.1 < Bi < 40$. This means that both external and internal resistance are operative here. The problem may be solved in one or two dimensions with appropriate boundary.

This solution is graphically equivalent to defining the boundaries of the cylinder assuming an intersection of an infinite slab and an infinite cylinder. That is, an infinite slab intersects an infinite cylinder to give the finite cylinder with the given dimensions (See the sketch below).

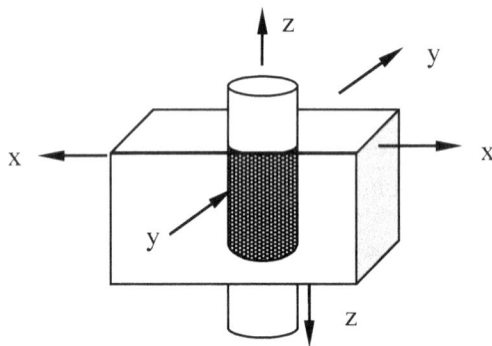

Thus

$$\left(\frac{T - T_\infty}{T_0 - T_\infty} \right)_{system} = \left(\frac{T - T_\infty}{T_0 - T_\infty} \right)_{slab} \times \left(\frac{T - T_\infty}{T_0 - T_\infty} \right)_{cylinder} \qquad (3)$$

Next, the Biot and Fourier numbers for the slab and cylinder are calculated as shown below.

For the Cylinder

$$\frac{k}{hR} = \frac{0.48 \ x \ 2}{2000 \ x \ 0.019} \cdot \frac{W}{mK} \frac{m^2 K}{Wm} = 0.025 \ (close \ to \ zero)$$

$$\frac{\alpha t}{R^2} = \frac{k}{\rho C p} \cdot \frac{t}{R^2} = \frac{0.48}{1025 \ x \ 4000} \cdot \frac{4 \ x \ t}{(0.019)^2} \cdot \frac{m^2}{s} \cdot \frac{s}{m^2} = 1.297 \ x \ 10^{-3} t$$

The minimum temperature to cook the carrot must be that at the centre of the carrot. That is, we are dealing with centreline temperatures, Tc, which is given as 365 K. This also means that $x = 0$ and $x/L = 0$.

For the Slab

$$\frac{k}{hL} = \frac{0.48}{2000 \ x \ 0.1} \cdot \frac{W}{mK} \frac{m^2 K}{Wm} = 0.0024 \ (close \ to \ zero)$$

$$\frac{\alpha t}{L^2} = \frac{k}{\rho C p} \cdot \frac{t}{L^2} = \frac{0.48}{1025 \ x \ 4000} \cdot \frac{t}{(0.1)^2} \cdot \frac{m^2}{s} \cdot \frac{s}{m^2} = 1.17 \ x \ 10^{-5} t$$

Again $\frac{x}{L} = 0$ but $\left(\frac{T-T_\infty}{T_0-T_\infty}\right)_{slab}$ is unknown. Hence from equation (3)

$$\left(\frac{T - T_\infty}{T_o - T_\infty}\right)_{system} = \frac{365 - 373}{295 - 373} = 0.103 \qquad (4)$$

That is

$$0.103 = \left(\frac{T - T_\infty}{T_o - T_\infty}\right)_{slab} x \left(\frac{T - T_\infty}{T_o - T_\infty}\right)_{cylinder} \qquad 5)$$

The solution procedure for the problem is to guess a value for t, use it to calculate the Biot and Fourier numbers. Use these calculated Biot and Fourier numbers to determine the Y for the slab and cylinder, respectively, from the Heisler chart. If the product of the Ys, as in equation (5), equals 0.103, then the guessed value of t is correct. Otherwise a new value of t is guessed until the correct product of the Ys is obtained.

For example, if we guess t = 500s, the table below illustrates the procedure.

	for the cylinder	for the slab
t, s	500	500
$\dfrac{\alpha t}{R^2}$	0.65	0.006
$\dfrac{hR}{k}$	0	0
$\dfrac{T - T_\infty}{T_0 - T_\infty}$	0.03	1.0

$$\left(\frac{T - T_\infty}{T_0 - T_\infty}\right)_{slab} x \left(\frac{T - T_\infty}{T_0 - T_\infty}\right)_{cylinder}$$
$$= 0.03$$

If we guess t = 350s, we get the table below

	for the cylinder	for the slab
t, s	350	350
$\dfrac{\alpha t}{R^2}$	0.454	0.004
$\dfrac{hR}{k}$	0	0

$$\frac{T - T_\infty}{T_0 - T_\infty} \qquad\qquad 0.1 \qquad\qquad 1.0$$

$$\left(\frac{T - T_\infty}{T_0 - T_\infty}\right)_{slab} x \left(\frac{T - T_\infty}{T_0 - T_\infty}\right)_{cylinder} = 0.1$$

Another guess may yield 0. 103 but this is not justified by the level of accuracy of interpolation in the charts, hence the time required is 350s or about 5.83 minutes. Ans

Example 8.06: Determine, from charts, the centreline temperature of a solid ceramic block 228mm x 228mm x 457mm initially at 302 K which was put in an oven, maintained at 950 C for 10 hours during which it sat with the 228mm x 457mm surface on an insulated surface. Assume that the convective heat transfer coefficient is 50 W/m²K and that the average thermal properties of the block during this process are k = 1.15 W/m.K; ρ = 3010 kg/m³; Cp = 837 J/kg.K

Answer

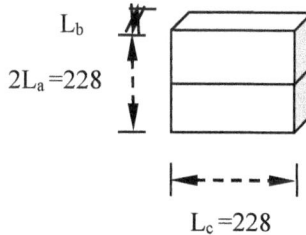

$$L_b$$
$$2L_a = 228$$
$$L_c = 228$$

The centreline temperature can be determined from charts so long as the numerical values of the chart parameters are obtained. These parameters are

$$m = \frac{k}{hL} = \frac{1}{Bi}; \quad n = \frac{x}{L}; \quad X = \frac{\alpha t}{L^2}; \quad Y = \frac{T_c - T_\infty}{T_0 - T_\infty}$$

Note that $Y = Y_a.Y_b.Y_c$ and that $n = 0$ refers to the centreline while $n = 1$ refers to the surface of the slab. Thus

		L_a	L_b	L_c
L, m		0.114	0.228	0.228
$m = \dfrac{k}{hL}$		0.202	0.1009	0.1009
$X = \dfrac{\alpha t}{L^2}$		1.263	0.316	0.316
$Y = \dfrac{T_c - T_\infty}{T_0 - T_\infty}$		0.15	0.67	0.67

Then
$$Y_{block} = Y_a.Y_b.Y_c = 0.15 \times 0.67 \times 0.67 = 0.067335$$

That is
$$\frac{T_c - T_\infty}{T_0 - T_\infty} = \frac{T_c - (950 + 273)}{302 - (950 + 273)} = 0.067335$$

which gives

$$T = 1160.98\ K \quad or \quad 887.98\ C.\ Ans$$

References For Chapter Eight

1   Carslaw, H.S and Jaeger, J.C., *Conduction of Heat in*
    Solids, Clarendon Press, Oxford, UK, 1959

2   Welty J.R: *Engineering Heat Transfer*, John Wiley and
    Sons New York, USA. 1978

3   Zill D. G. and Cullen M. R., *Differential Equations with*
    Boundary Value Problems; Brooks/Cole Publishing Co.,
    California, USA, 1997

APPENDIX I: TYPICAL HEISLER CHARTS
(Wikipedia, 2011)

Fig. AI. 1a: Temperature Distribution in a Plane Wall of Thickness 2L (Wikipedia, 2011)

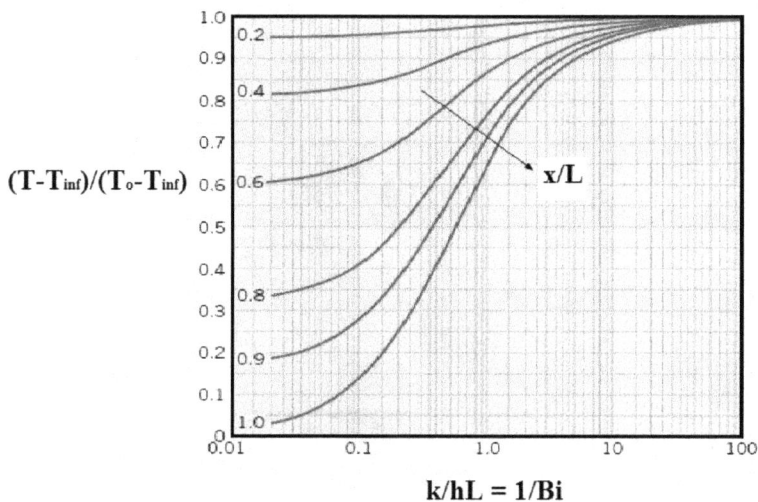

$(T-T_{inf})/(T_o-T_{inf})$

x/L

$k/hL = 1/Bi$

Fig. AI. 2a: Temperature Distribution in an infinite Cylinder of radius r_o

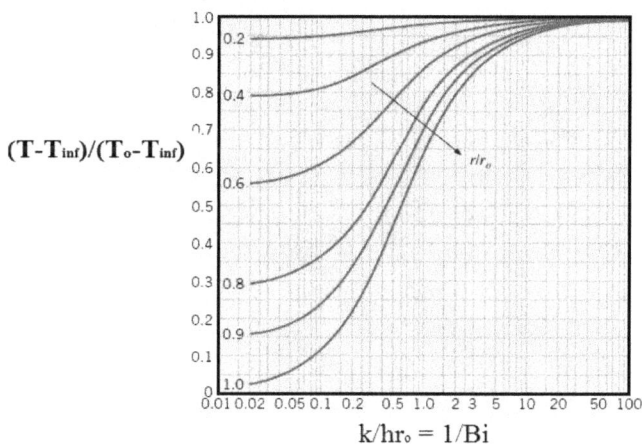

$(T-T_{inf})/(T_o-T_{inf})$

r/r_o

$k/hr_o = 1/Bi$

227

For the infinitely long cylinder, the Heisler chart is based on the first term in an exact solution to a Bessel function.

Fig. AI. 3a: Temperature Distribution in a Sphere of radius ro

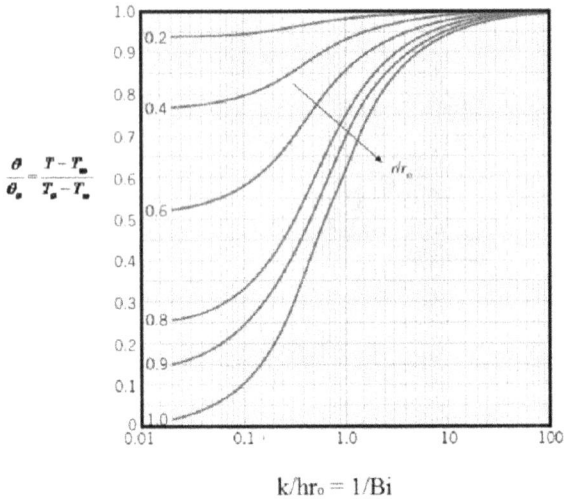

$$\frac{\theta}{\theta_o} = \frac{T - T_\infty}{T_o - T_\infty}$$

$$k/hr_o = 1/Bi$$

The Heisler Chart for a sphere is based on the first term in the exact Fourier series solution.

Fig. AI. 2c: Centreline Temperatures as a Function of Time for an infinite Cylinder of radius r_o

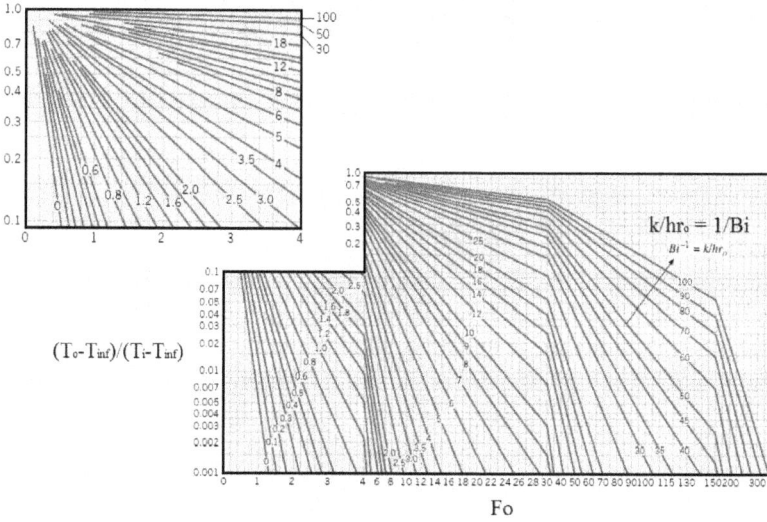

Fig. AI. 3c: Centre Temperature as a Function of Time in a Sphere of radius r_o

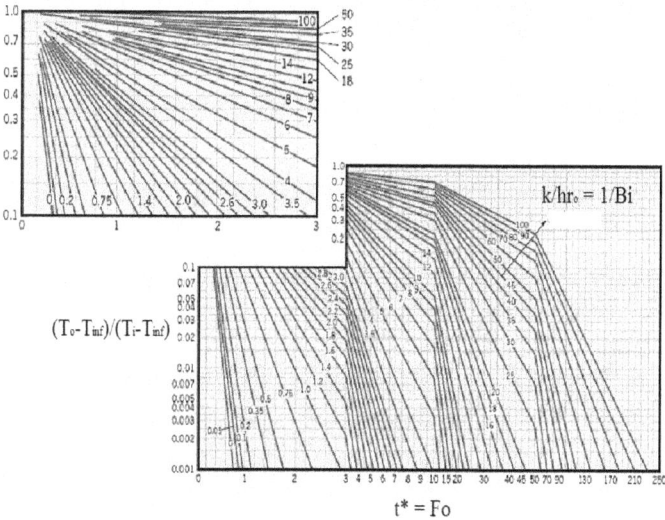

APPENDIX II
THERMAL CONDUCTIVITY OF SOME METALS
(Engineering Toolbox, 2009)

Metal	Temperature, t, C	Thermal Conductivity, k, W/mK
Admiralty Brass	20.00	110.78
Aluminium, pure	20.00	204.26
	93.33	214.64
	204.44	249.26
Aluminium Bronze	20.00	76.16
Antimony	20.00	18.52
Beryllium	20.00	218.11
Beryllium Copper	20.00	65.78
Bismuth	20.00	8.48
Cadmium	20.00	93.47
Carbon Steel, max	20.00	53.66
Carbon Steel, max 1.5% C	20.00	36.35
	400.00	32.89
	1200.00	29.43
Cast Iron, grey	21.11	46.7 -79.6
Chromium	20.00	90.01
Cobalt	20.00	69.24
Copper, pure	20.00	386.01
	300.00	368.70
	600.00	353.12

Copper bronze (75% Cu, 25% Zi)	20.00	25.97
Copper brass (70% Cu, 30% Zi)	20.00	110.78
Cupronickel	20.00	29.43
Gold	20.00	315.04
Hastelloy B	-17.78	10.39
Hastelloy C	21.11	8.66
Inconel	21.11 - 100	14.54
Incoloy	0 - 100	11.77
Iridium	20.00	147.14
Iron, nodular pearlitic	100.00	31.16
Iron, pure	20.00	72.70
	300.00	55.39
	1000.00	34.62
Iron, wrought	20.00	58.85
Lead	20.00	34.62
	300.00	29.77
Manganese Bronze	20.00	105.59
Magnesium	20.00	159.08
Mercury	20.00	8.40
Molybdenum	20.00	140.21
Monel	0 - 100	25.97

Nickel	20.00	90.01
Nickel Wrought	0 - 100	60.6 - 90.0
Niobium (Columbium)	20.00	51.93
Osmium	20.00	60.59
Platinum	20.00	72.70
Plutonium	20.00	7.96
Potassium	20.00	100.05
Red Brass	20.00	159.25
Rhodium	20.00	150.08
Selenium	20.00	0.52
Silicon	20.00	83.61
Silver, pure	20.00	406.79
Sodium	20.00	134.15
Stainless Steel	20.00	12.1 – 45.0
Tantalum	20.00	53.66
Thorium	20.00	41.54
Tin	0.00	62.3 - 67.5
Titanium	20.00	19.0 – 22.5
Tungsten	20.00	162.7 - 173.1
Uranium	20.00	24.23
Vanadium	20.00	60.59
Wrought Carbon Steel	0.00	58.85
Yellow Brass	20.00	115.98

Zinc		115.98	
Zirconium		251.00	

APPENDIX III
THERMAL CONDUCTIVITY OF SOME COMMON MATERIALS
(Engineering Toolbox, 2009)

Material/Substance	**Thermal Conductivity - k - (W/m . K)**		
	Temperature (° C)		
	25	125	225
Acetone	0.16		
Acrylic	0.2		
Air	0.024		
Alcohol	0.17		
Aluminium	250	255	250
Aluminium Oxide	30		
Ammonia	0.022		
Antimony	18.5		
Argon	0.016		
Asbestos-cement board	0.744		
Asbestos-cement sheets	0.166		

Asbestos-cement	2.07
Asbestos, loosely packed	0.15
Asbestos mill board	0.14
Asphalt	0.75
Balsa	0.048
Bitumen	0.17
Benzene	0.16
Beryllium	218
Brass	109
Brick dense	1.31
Brick work	0.69
Cadmium	92
Carbon	1.7
Carbon dioxide	0.0146
Cement, Portland	0.29
Cement, mortar	1.73
Chalk	0.09
Chrome Nickel Steel	16.3
Clay, dry to moist	0.15 - 1.8

Clay, saturated	0.6 - 2.5		
Cobalt	69		
Concrete, light	0.42		
Concrete, stone	1.7		
Constantan	22		
Copper	401	400	398
Corian (ceramic filled)	1.06		
Corkboard	0.043		
Cork, re-granulated	0.044		
Cork	0.07		
Cotton	0.03		
Carbon Steel	54	51	47
Cotton Wool insulation	0.029		
Diatomaceous earth (Sil-o-cel)	0.06		
Earth, dry	1.5		
Ether	0.14		
Epoxy	0.35		
Felt insulation	0.04		
Fibreglass	0.04		

Fibre insulating board	0.048		
Fibre hardboard	0.2		
Fireclay brick 500 °C	1.4		
Foam glass	0.045		
Freon 12	0.073		
Gasoline	0.15		
Glass	1.05		
Glass, Pearls, dry	0.18		
Glass, Pearls, saturated	0.76		
Glass, window	0.96		
Glass, wool Insulation	0.04		
Glycerol	0.28		
Gold	310	312	310
Granite	1.7 - 4.0		
Gypsum or plaster board	0.17		
Hair felt	0.05		
Hardboard high density	0.15		
Hardwoods (oak, maple)	0.16		
Helium	0.142		

Hydrogen	0.168		
Ice (0oC, 32oF)	2.18		
Insulation materials	0.035 - 0.16		
Iridium	147		
Iron	80	68	60
Iron, wrought	59		
Iron, cast	55		
Kapok insulation	0.034		
Kerosene	0.15		
Lead (Pb)	35		
Leather, dry	0.14		
Limestone	1.26 - 1.33		
Magnesia insulation	0.07		
Magnetite	4.15		
Magnesium	156		
Marble	2.08 - 2.94		
Mercury	8		
Methane	0.030		
Methanol	0.21		

Mica	0.71
Mineral insulation materials,	
Molybdenum	138
Monel	26
Nickel	91
Nitrogen	0.024
Nylon 6	0.25
Oil, machine lubricating SAE 50	0.15
Olive oil	0.17
Oxygen	0.024
Paper	0.05
Paraffin Wax	0.25
Perlite, atmospheric pressure	0.031
Perlite, vacuum	0.00137
Plaster, gypsum	0.48
Plaster, metal lath	0.47
Plaster, wood lath	0.28
Plastics, foamed	
Plastics, solid	

Platinum	70	71	72
Plywood	0.13		
Polyethylene HD	0.42 - 0.51		
Polypropylene	0.1 - 0.22		
Polystyrene expanded	0.03		
Porcelain	1.5		
PTFE	0.25		
PVC	0.19		
Pyrex glass	1.005		
Quartz mineral	3		
Rock, solid	2 - 7		
Rock, porous volcanic (Tuff)	0.5 - 2.5		
Rock Wool insulation	0.045		
Sand, dry	0.15 - 0.25		
Sand, moist	0.25 - 2		
Sand, saturated	2 - 4		
Sandstone	1.7		
Sawdust	0.08		
Silica aerogel	0.02		

Silicone oil	0.1		
Silver	429		
Snow (temp < 0 °C)	0.05 - 0.25		
Sodium	84		
Softwoods (fir, pine ..)	0.12		
Soil, with organic matter	0.15 - 2		
Soil, saturated	0.6 - 4		
Steel, Carbon 1%	43		
Stainless Steel	16	17	19
Straw insulation	0.09		
Styrofoam	0.033		
Tin (Sn)	67		
Zinc (Zn)	116		
Urethane foam	0.021		
Vermiculite	0.058		
Vinyl ester	0.25		
Water	0.58		
Water, vapour (steam)		0.016	
Wood across the grain, white pine	0.12		

Wood across the grain, balsa	0.055		
Wood across the grain, yellow pine	0.147		
Wood, oak	0.17		
Wool, felt	0.07		

www.ingramcontent.com/pod-product-compliance
Lightning Source LLC
Chambersburg PA
CBHW071632200326

41519CB00012BA/2263